WILLIAM CONGREVE

A TREATISE

ON THE

GENERAL PRINCIPLES, POWERS AND FACILITY

OF APPLICATION

OF THE

CONGREVE ROCKET SYSTEM

AS COMPARED WITH ARTILLERY

Elibron Classics
www.elibron.com

Elibron Classics series.

ISBN 1-4021-7960-X (paperback)
ISBN 1-4212-8738-2 (hardcover)

This Elibron Classics Replica Edition is an unabridged facsimile of the edition published in 1827 by Longman, Rees, Orme, Brown, and Green, London.

A TREATISE

ON THE

GENERAL PRINCIPLES, POWERS, AND FACILITY OF APPLICATION

OF THE

CONGREVE ROCKET SYSTEM,

AS COMPARED WITH ARTILLERY:

SHOWING

THE VARIOUS APPLICATIONS OF THIS WEAPON,

BOTH FOR SEA AND LAND SERVICE,

AND ITS DIFFERENT USES IN THE FIELD AND IN SIEGES.

ILLUSTRATED BY

PLATES OF THE PRINCIPAL EXERCISES AND CASES OF ACTUAL SERVICE.

WITH

A DEMONSTRATION

OF

THE COMPARATIVE ECONOMY OF THE SYSTEM.

BY

MAJOR-GEN. SIR W. CONGREVE, BART. M.P.

F.R.S., &c. &c. &c.

AND DEDICATED TO HIS MOST GRACIOUS MAJESTY, GEORGE THE FOURTH.

LONDON:

PRINTED FOR LONGMAN, REES, ORME, BROWN, AND GREEN,

PATERNOSTER-ROW.

1827.

London:
Printed by A. & R. Spottiswoode,
New-Street-Square.

TO

HIS MOST GRACIOUS MAJESTY,

GEORGE THE FOURTH.

SIRE,

MAY it please Your Majesty graciously to permit me to dedicate to Your Majesty the following pages, explaining the present powers of the Rocket System, with the various improvements which I have made therein since it was first honoured by Your Majesty's protection and patronage. It is now upwards of twenty years since I first had the honour of laying at Your Majesty's feet a plan which I had devised for a maritime attack on the enemy's sea-port by means of Rockets. Your Majesty having been pleased to approve and countenance this plan, it was immediately carried into effect, and since that period an expedition has scarcely sailed from this country in which Rockets have not been largely employed. During this period various improvements have been made in the weapon, and although it was at first merely contemplated as applicable to naval warfare, it has since been extended to every branch of the land service, and especially as an arm for cavalry. In its application to this part of the service, Your Majesty has been pleased to feel the greatest interest; an interest which has been fully justified by the important results attained at the battle of Leipsic, by a small detachment of the British horse artillery, which were then armed as Rocketeers, and who were, indeed, the only British troops engaged at that memorable battle.

Your Majesty has been graciously pleased at different times to take into consideration the details of these various improvements; and as the whole system has now received a consistent and, it may be hoped, a permanent form, I have thought it my duty to collect all its parts, and arrange them in the form of a treatise, which, while I trust it will give to the British army that confidence in this new weapon to which it is entitled, will, at the same time, prove to the world that the Rocket is not unworthy of the high and distinguished patronage which Your Majesty has from the first been pleased to afford it.

I have the honour to subscribe myself,

With the utmost gratitude and devotion,

Sire,

Your Majesty's most faithful, dutiful,

And obedient Servant and Subject,

WILLIAM CONGREVE.

August, 1827.

ADVERTISEMENT.

It is an extraordinary circumstance, that the powers of the Rocket are more appreciated on the Continent than in this country, and that greater efforts are making in the formation of Rocket Establishments by foreign powers, than in the preservation and extension of the use of this arm in our own country. This can only be attributed to the little information that prevails here on the subject: I have therefore thought it my duty to draw up the following general statement of the capability of the weapon, in its present state of perfection, that my own countrymen may be more generally informed of its importance, and that we, who are the original inventors, may not lose the advantage of it from an ignorance of its value.

What I have here published goes merely to the principle of its tactics, and of the different manœuvres of which it is capable in the field, which have not only been witnessed in action by our allies, but by such of our enemies as it has been employed against.

From this work, therefore, those enemies can derive no advantage, since no disclosure whatever is in any part made, as to the construction of the weapon itself, while I do trust, that the general knowledge which the British army may thus obtain, of the benefits which every branch of the service might derive from the full extension and due application of this arm, will create an interest in its general adoption which may lead to the best results in future campaigns.

I cannot conclude this address to the public, without stating that the Rocket System not only owes its existence but its present state of perfection to the patronage of His Majesty.

My original plan for the attack on Boulogne, was first presented in the year 1805, at the pavilion at Brighton, and was so far honoured by the approbation of His Majesty, then Prince of Wales, that His Royal Highness wrote a letter in his own hand to Mr. Pitt, recommending his immediate attention to its adoption, and sent me round to Walmer Castle (at which place Mr. Pitt then was) in the cutter which attended His Royal Highness at Brighton, that no time might be lost. The result was, that preparations were immediately commenced for the attack, which took place in the autumn of that year.

The same distinguished support and recommendation, induced the renewal of the attack on Boulogne in the following year, on Lord Moira becoming Master-General of the Ordnance.

And, lastly, it was by the direct order of His Majesty, then Prince Regent, that the Rocket troop was sent under the command of Captain Bogue to join the allied armies, previous to the battle of Leipsic, where this little band (the only Englishmen in that battle) had the happiness to prove themselves, and the weapon they wielded, worthy of the distinguished patronage which they had received from their Royal Master. This was the first application that had ever been made of Rockets to the service of cavalry. And here it ought to be stated, that this troop was actually formed under the immediate direction of the Prince

Regent; and also that the most important parts of the detail were planned by His Royal Highness.

Duty and gratitude, therefore, to my Royal Master, for all the favours and reward which I have received, as the inventor of this weapon, even if the other motives for the public good which have already been stated did not exist, would call upon me, while my health still permits me, to prove by the following pages the importance of an invention which has been thus so highly distinguished by the Royal Patronage.

The Rocket possesses two important peculiarities in its application to the different branches of the service. The first and perhaps most important of these is, its capability of being used, to a powerful and almost incredible extent, by cavalry, *without wheel carriages at all*, so that, wherever horses can travel, this species of artillery may be brought into action without incumbrance, each dragoon carrying in holsters four rounds of 6-pounder ammunition, ranging as far, and with equal force, as a 6-pounder shot; while every third man carries his tube or rocket gun, slung like a carbine, and not much heavier, yet capable of discharging this 6-pounder ammunition with quite as much precision as is necessary in general service against the enemy. The Rocket troop is told off in sections of threes, each section being a complete independent force armed with its 6-pounder Rocket gun, and 12 rounds of 6-pounder ammunition, without the incumbrance of wheel-carriages; thus 1000 men so armed would carry into action a battery of upwards of three hundred 6-pounder bouches à feu, and 4000 rounds of this most formidable ammunition.

The vast importance of thus giving to cavalry the power of artillery, *without materially interfering with any of its other duties,*

I need not dwell upon, even supposing that, instead of arming whole regiments in this way, which would enable them to carry every thing before them, only two squadrons were so armed in each regiment.

The second peculiarity of this arm is, that when wheel carriages are used, from the extreme lightness of the cylinder required for discharging the rocket, as compared with the weight of common field ordnance (the light 6-pounder gun weighing six hundred weight, while the tube required to project the 6-pounder rocket to the same distance, and with the same force, weighs only twenty pounds), the number of bouches à feu on each carriage may be increased in proportion to the calibre of the rocket, thereby making every rocket carriage a *volley* carriage, instead of arming it with a single bouche à feu. Thus a 24-pounder carriage may be constructed not heavier than the lightest 6-pounder horse-artillery carriage, capable of firing four 24-pounder rockets in a volley; an 18-pounder carriage may in like manner be constructed to fire six 18-pounders in a volley; a 12-pounder, eight 12-pounders; a 9-pounder, ten 9-pounders; a 6-pounder, twelve 6-pounders; and a 3-pounder capable of firing twenty 3-pounder rockets in a volley.

Unfortunately, however, under the prevailing idea that the rocket should only be used as an auxiliary to artillery, these important peculiarities of power seem almost entirely to have been overlooked. The principal plan of employing rockets now contemplated at Woolwich is to *limit the conveyance of them in action, to heavy ammunition waggons*, the same as those used for the transport of common artillery ammunition, by which means the facility of their application is literally brought below that of the heaviest field artillery. The reason assigned for this, is to avoid the multiplication of different constructions of carriages. Whether indeed the rocket is to be considered as an *auxiliary*, or as a *principal*, is a matter of very little consequence; but, in whatever class it may be ranked, all I contend for is, that the utmost may be made of its peculiar powers and properties. That a reserve of rockets should be maintained in waggons as well as of artillery ammunition, there can be no question; but it is quite absurd to say, that, when this arm is to be carried into action by wheel carriages, carriages appropriate to the powers and properties of the rocket should not be constructed, as well as for every other species of artillery.

WILLIAM CONGREVE.

INTRODUCTION.

SHOWING THE ORIGIN, PROGRESS, AND GENERAL PRINCIPLES OF THE ROCKET SYSTEM, FOR SEA AND LAND SERVICE.

SECTION I.

OF THE ORIGIN AND PROGRESS OF THE INVENTION.

In the year 1804 it first occurred to me, that, as the projectile force of the rocket is exerted without any re-action upon the point from which it is discharged, it might be successfully applied, both afloat and ashore, as a military engine, in many cases where, from the violent recoil produced by exploding gunpowder, the use of artillery, according to the ordinary modes, is either entirely impossible, or at best very limited. But the desideratum and the difficulty were to obtain, upon the principle of the rocket, a force sufficient to render the projectile, thus to be conveyed, of sufficient importance either as to range or quantity of matter projected. I knew that rockets were used for military purposes in India; but that their magnitude was inconsiderable, and their range not exceeding 1000 yards. I knew, also, that some years since several experiments had been made in the royal laboratory by General Desaguliers, then fire-master, for the construction of large rockets; but that they had not succeeded, and that very few of them would even rise off the stand. Still, however, I could not but believe that the force of gunpowder was capable of due modification in this way as well as in the ordinary mode, both for length of range, and for the conveyance of considerable weights, and therefore felt determined to convince myself by actual experiment.

The first point which I resolved to ascertain, was as to the range. I immediately procured, at my own cost, the largest rockets I could get made in

London, but found on trial that they were none of them capable of more than five or six hundred yards range, according to the usual modes of construction. Various plans suggested themselves to my mind, and were successively tried. It is not necessary, however, to enter into the detail of them here; it is sufficient to state that I obtained, at length, ranges of 1500 yards from the same class of rocket which at first would only range 600. I found, also, that as far as I had been able to proceed with the rockets made in London, the ranges increased in a certain proportion to their size. This, therefore, was sufficient to convince me that there did actually exist, in this principle of projectile force, a power capable of great extension, both as to range and weight. Hitherto I had laboured at my own cost, and had expended several hundred pounds in different experiments; but as I had great hopes in the construction of larger rockets, and in the better workmanship of the royal laboratory, I applied to Lord Chatham for permission to have some large rockets made at Woolwich, and his Lordship was pleased to grant my request.

In the first instance, therefore, I desired to have several 6-pounder rockets constructed according to the laboratory proportions. On trial, they would scarcely range 600 yards, though the same-sized rocket was very shortly brought, by being constructed on the principles which I had previously ascertained, to range full 2000 yards.

In this state of the weapon, I resolved to lay before Mr. Pitt a plan for the annoyance of Boulogne by fire rockets; in consequence of which, he was pleased to attend an experiment, with Lord Castlereagh and Lord Mulgrave, at Woolwich; when, from their conviction of the powers of the weapon, it was determined that an attack should be made on the flotilla by projecting a large quantity from men of war's launches. Sir Sidney Smith was appointed to the command of the Boulogne squadron, and ten launches were fitted under my direction for this particular service. Unfortunately it was so late in the season when the preparations were commenced, that they could not be completed till the middle of November. On the 18th of that month, 1805, the squadron and boats were collected off Boulogne. The attack was fixed for the night of the 21st — the weather had been favourable the whole of the day; I had arranged every thing during the morning in the different launches; and was proceeding in with the boats, when about eight o'clock the wind suddenly shifted to the N.W. and blew with such violence, that Sir Sidney was compelled to recal the vessels without a rocket having been fired. In fine, it was with difficulty the squadron got out of the bay; and so much did the gale increase during the night that five of the ten launches were swamped. It was thus impossible to renew the attack that year.

In the spring of 1806, I conceived the plan, and obtained permission to proceed in making still larger rockets, and forming them with iron cases instead of paper. The event answered my utmost expectations: I succeeded in constructing rockets weighing thirty-two pounds, capable of conveying as much carcass composition as a ten-inch spherical carcass, and which would, on an average, range 3,000 yards. I discovered also, that the great length of the stick, given by the laboratory proportions, was by no means necessary to ranging rockets, and reduced that length from 25 feet to 15 feet, making the 32-pounder stick even less than that of the ordinary 18-pounder, which gave vast additional facility for service. These improvements induced me to urge a renewal of the attack on Boulogne; and, in consequence of my solicitations, Lord Moira, who had then just come to the Ordnance, and Lord Howick, the first lord of the Admiralty, thought proper to attend an experiment at Woolwich. My only desire was that they should see the weapon, and decide from their own conviction of its powers — they did so, and the result of this conviction was an order to me for the immediate construction of a quantity of rockets such as they had seen; and a special warrant of Admiralty, empowering me to superintend such preparations as might be necessary for the attack which I had proposed, and which was ordered to be carried into effect under the direction of Commodore Owen.

But the second attempt was not fated to be attended with circumstances much more favourable than the first. At the moment when the preparations were complete, and in the best season of the year, a negotiation for peace was entered upon, and the embarkation of Lord Lauderdale for France was the signal for the suspension of operations. Thus again was the summer lost; nor was it until the departure of Lord Lauderdale from Paris that the attack took place. On the 8th of October, the squadron was off Boulogne. — Lord Lauderdale was on his way to England, but as the wind suited, Commodore Owen resolved not to delay it any longer. Accordingly, on the evening of that day, eighteen of the boats of the squadron took their frames and rockets on board, and rowed into the bay.

The attack was made with as much order as could be expected in so new a service. In about half an hour above 200 rockets were discharged. The dismay and astonishment of the enemy were complete — not a shot was returned — and in less than ten minutes after the first discharge, the town was discovered to be on fire. What was the extent of the mischief effected cannot be known, though it is to be feared the basin escaped without injury, from the rockets being thrown too much to the left: it was reported, however, that some vessels in the harbour had been destroyed, and it is certain a considerable range of buildings, supposed to be barracks or storehouses, were burnt. The fire indeed could not, from its duration, be trifling, having lasted from two o'clock in the morning to the next evening. The ruins of eight buildings were distinctly to be counted from the Clyde; and there is every reason to suppose the damage must have been much more extensive, from the extreme caution with which Lord Lauderdale and his suite were guarded in their passage through the town a few days afterwards, not one of them having been suffered to leave the inn at which they were placed, nor any one permitted to have communication with them; even in passing through the streets, they were conveyed in close cabriolets.

This jealousy on the part of the enemy was a sufficient proof that more had been effected by the attack than met the eye of those on board the squadron: the duration of the fire, however, and the ruins that could be seen, showed that the enemy had suffered considerably; it was only to be regretted that the conflagration had not taken effect more to the right, where the bulk of the flotilla lay: nevertheless, the proof of the efficiency of the weapon, and the liability of Boulogne to a repetition of the attack, were certainly considered complete; since it could not be doubted that what had destroyed houses would have destroyed the shipping, had it fallen amongst them; and it is certain that the part of the town burnt was more remote from the boats employed to throw the rockets than the flotilla itself was — at the same time, also, that the facility of using this weapon afloat was fully demonstrated.

In the year 1807 Rockets were again brought into service at Copenhagen, under my direction. In 1809 I went to Basque roads with a large equipment of Rockets, on which occasion 1200 of them were distributed in different parts of the rigging of the fire-ships, as described in plate 12.; and in the same year I joined the expedition to Walcheren also, with a large equipment under my direction. Subsequently to this, a Rocket troop was formed at Woolwich, under the command of Captain Bogue. This troop joined the great allied army on the Continent, and was present at the battle of Leipsic where it greatly distinguished itself. The Rocketeers were the only English troops present at that ever-memorable battle, and are acknowledged to have rendered signal service; for which His Majesty has been graciously pleased to grant them permission to wear the word "Leipsic" on their helmets and shabraques. Captain Bogue was unfortunately killed at this battle. The Rockets were but seldom brought into action in the peninsula, but are acknowledged to have saved a brigade of guards in the passage of the Adour.

I have also had the satisfaction of introducing this weapon into the service of the East India Company, and for that purpose, in 1817, I established a manufactory from which Rockets have from time to time been sent to India, with instructions for the formation of different corps for using them. An officer of the India Company's service having, however, succeeded to a certain degree in imitating my Rockets, at all events, in their external appearance, I understand the Government in India have established a factory for their manufacture under the direction of this officer; although I myself have hitherto received no remuneration for my exertions in the service of the Company, and am, indeed, a considerable loser by the establishment I have erected for their supply.

SECTION II.

OF THE USE AND ADVANTAGE OF THE ROCKET SYSTEM AS APPLIED TO NAVAL PURPOSES.

It is in the application of the Rocket to naval purposes that the advantages to be derived from the first principle of its flight are most conspicuous; for it is the power it possesses of projecting itself without re-action upon the point whence it departs, that enables us by means of the Rocket to discharge, from the smallest boats in the navy, carcasses of the same force as can only be thrown by the ordinary means from vessels of very considerable tonnage, and to project the one in vastly greater quantities than the other, even under circumstances of the greatest disparity in the dimensions of the vessels employed; so that, by means of the rocket, as much may be accomplished in a few minutes by a few small vessels, as, in the ordinary mode of naval bombardment, would be the work of hours, with the same number of the largest bomb ships.

Thus it is possible to equip a gun-brig so as to discharge twenty 32-pounder rockets in a broadside, firing them as quick as may be desired in succession. Ten gun-brigs so equipped, therefore, would, in the space of a few minutes, discharge 200 such rockets, each equal, or rather, as I believe we have proved, superior to a 10-inch spherical carcass; whereas ten ships would not throw the same number of spherical carcasses in less than two hours; during which time the gun-brigs would be able to repeat their broadsides at least five or six times, throwing in that time not less than 1000 or 1200 rockets, which convey *nearly four tons* of combustible matter.

For a naval attack, therefore, of any strongly-fortified town or harbour, where the object is the conflagration of the town or shipping, it surely is not presuming too much to say, if the reasoning on the powers of the rocket in Section III. be admitted, that there is no weapon at present known which possesses the same powers.

But there is even a stronger case which may be put; where, for instance, it is desirable, for the sake of security, and on account of any particularly formidable defence on the part of the enemy, to perform any such service, *absolutely* by a coup de main, and in the shortest possible time—20 gun-brigs would discharge, *in one broadside*, 400 rockets, and this number would make an impression on any place; nor is there any thing difficult in supposing the employment of 20 gun-brigs in such an exploit; whereas, to accomplish the same thing by mortars in the same time, would require no less than two hundred mortar vessels, a number totally impracticable on all accounts. For services of this description, I do not, therefore, hesitate to affirm, that were this weapon properly put into the hands of the navy, its importance would be extreme.

But, in fact, the advantages of it for naval purposes do not rest here; for such is the simplicity of the system, that the equipment required for the discharge of these broadsides consists principally of the spare spars of the vessel—the vessel itself requires no additional work, no change of construction; the apparatus is of the simplest kind, and interferes not with any other duty the vessel may have to perform. In short, every vessel, every boat in the navy, becomes capable of the powers of the bomb ship, as far as the throwing carcasses is concerned, by having only a few boxes of rockets and a frame put on board her.

The saving, therefore, that attaches to this mode of throwing carcasses afloat, instead of the ordinary method where the vessel is expensive, is constructed on purpose, and fit for little else, is almost incalculable. It is obvious that the late introduction of steam navigation opens a new field for naval warfare—by its combination with heavy rockets. The necessity of confining the dimensions of steam vessels to ships of a light draught of water must prescribe very narrow limits to the application of heavy ordnance in such vessels; while the simplicity of the rigging and the clearness of the deck of a steam vessel from cordage, and from the number of men required for trimming sails, render vessels worked by steam peculiarly applicable to the most extensive and powerful employment of heavy Rockets of all descriptions.

SECTION III.

OF THE USE AND ADVANTAGE OF THE ROCKET SYSTEM, AS APPLIED TO THE LAND SERVICE.

As in the application of the Rocket to *naval* purposes, that principle of its motion by which it exerts no re-action on its fulcrum is found to constitute the basis of all its advantages, so shall we find, in its application to *military* purposes, that the inherent power of flight which it possesses, without the incumbrance of ordnance, is the foundation of its peculiar facilities for this latter branch of the service.

In this section, therefore, I shall first consider the application of the weapon in sieges; and, in the second place, what probable advantage may be derived from it in the field. But before we proceed in its application to bombardments, we must first consider how far the operation of carcasses in general is of importance in them.

It appears to me that bombardments are of three kinds — first, the bombardment of any arsenal, or magazines, where the object is solely that of burning them — in which the carcass is not only the most important weapon, but perhaps is exclusively efficient: — secondly, the bombardment of large fortified towns, where the garrison forms an inconsiderable part of the population, in which, though the use of shells is greatly important, still that of carcasses will be found much more so: — and, thirdly, the bombardment of mere fortresses, where the garrison is the great consideration — in which I consider the shell as the principal weapon, and the carcass of very little use.

To demonstrate the first of these positions requires but little argument, as where the sole object is conflagration, those means only are required which tend to that effect — and the best that can be used are evidently those which contribute to it most certainly.

The second position requires a little more consideration. I contend, however, that, in the attack of populous cities, the burning of the town and the destruction of houses and property is of more importance, and will lead, if carried to any great extreme, more speedily to a surrender than the killing of the garrison, and the throwing shells into the town — for the loss of lives, occasioned by the throwing shells amongst the garrison, is but little felt —

and the loss of lives amongst the populace need not be felt at all, if they desert their houses, and retire, as they may, in a large city, to a place of safety: — the populace can only, therefore, be seriously injured by the loss of their property* — let them but once behold their city extensively in flames — let them see the fire destroying all before it — let them feel that, to save their lives, they must abandon every thing to destruction — that, under these circumstances, there are no limits to the conflagration — but that the devouring element will soon follow them, where the shells and bullets of the enemy cannot — and what is the result? — that the inhabitants, who, in such towns, are more powerful than the garrison, will compel the garrison to capitulate. Why, therefore, should we hesitate to say, that, in this case, the importance of the shell is less than that of the carcass?

We now come, however, to that position where exactly the reverse is true — and as it depends on nearly the same principles, little is required to prove it. In fortresses, or small fortified towns, the garrison is every thing — it is but of little importance to them that the few streets contained within their work are burnt, nor will they surrender until their defences are destroyed, and their bomb proofs shattered.

Thus in two out of three of the cases of bombardment, the use of carcasses is pre-eminent. We have now, therefore, to enquire what advantages peculiarly accrue from the adoption of the rocket carcass in such bombardments — and here the most cursory view of the subject presents us with an immense saving of time, of labour, and of expense — the conveyance of heavy mortars by sea, their embarkation and disembarkation, are works of sufficient difficulty — but their carriage by land is ever most tedious and expensive, and in frequent situations of difficult passes or morasses, totally impracticable. Surely, therefore, there can be no question as to the accumulated advantages of substituting for a weapon, which, to be made effectual, requires a considerable number of these ponderous machines, one, the burning powers of which are proved to be at least equal, and which may, without difficulty, if

* Tremendous as are the effects of shells amongst houses, still it appears to me, that a town may be more expeditiously destroyed by powerful carcasses than by shells; because, allowing that the shell does not often set fire, it requires a vast number of them to destroy any great number of houses; whereas a few powerful vollies of carcasses produce an extensive conflagration — and such a fire may be fed by *occasional* discharges until the whole is destroyed. Here, therefore, it appears that the great use of shells is to prevent the people from extinguishing the conflagration.

required, be conveyed by the soldiers themselves into any situation where it may be used even without additional apparatus. In fact, the rocket requires only a slope * of a very few yards in length, to fire it in very large quantities; and saves, therefore, in addition to the labour of the conveyance of the mortars, the lingering process of making large batteries, and of laying numerous platforms — nor indeed are regular approaches necessary for the use of the rocket carcass; for so simple are the means used to project it, that they need not be thrown twice from the same place — and so rapidly would the business be performed, in each situation, that the enemy need never be aware of the quarter from which they were about to be thrown, and consequently his fire need be but little dreaded — while no difficulty of ground, no acclivity nor morass, would prevent the application of it — either to the destruction of arsenals, where it appears that it would alone be sufficient — or to the taking of towns, where it appears at least to be capable of reducing half the number of mortars required, and where the smallest reduction of their number is of so much consequence to the service on every account.

What, then, in all these comparisons, becomes of the objections that have been made to the expense of the rocket carcass, when it will be found, that, on comparing the mere projectiles, the rocket is absolutely cheaper than the spherical carcass, and when there is scarcely any additional expense attached to the former; whereas the latter cannot be employed without the almost incalculable cost of mortars, mortar beds, platforms, batteries, and magazines, with all the carriages and horses necessary to drag them; independent of the conveyance by water, and of the great comparative difference in the number of men required to use the one and the other in action.

There is another point of the greatest advantage attaching to the rocket in the attack of arsenals and towns, which has not yet been noticed, and which hinges entirely on the great difference of time between the execution of such a service by rockets, and by the ordinary means. There are many situations where such an enterprise might be accomplished by the landing of troops possessing the means of rapid bombardment, which could never be thought of if the slow process of landing mortars, &c. were to be resorted to.

But I do not conceive the use of the weapon to be confined entirely to

* All that is required to discharge the rocket is a bank to lay it against, sloped to the proper elevation, and scored down to receive and direct it — from such a trench, 30 yards long, 100 rockets may be fired in a volley.

the attack of towns—it appears to me that it is capable of considerable service in their defence, and that especially where fascines and frame-work are used in the attack—besides which, in any number of rockets fired at low angles, many would continually get into the trenches—and such is the nature of the weapon, that having once found its way in, it would constantly proceed in the direction of the trench, however crooked, until its force were expended—so that the rocket would, in some measure, afford an enfilading fire, and a very destructive one, even for the most indirect approach.

I now propose to devote a few lines to the use and advantages which I conceive may be made of rockets in the field. In the first place, therefore, field carriages may be constructed of the lightest kind, (for they have only the weight of the ammunition to carry, no ordnance, nor any re-action to support,) to be attached to every park of artillery, for the purpose either of using the rocket carcass in such regular sieges as may occur during a campaign, or for the occasional business of destroying magazines, of disposting an enemy by the burning of villages, or for the annoyance of camps, all of which might thus be accomplished with the lightest description of artillery, yet at such ranges as at present are unknown in the field, even by the heaviest field ordnance.

But in addition to the use of the rocket carcass in the field, it must here be stated, that the powers of the 32-pounder rocket are such, that, being laid on the surface of tolerably smooth ground, it will carry a 24-pound from 1200 to 1500 yards.

It is therefore capable of bringing into action, against troops in the field, a much heavier description of ammunition than can now be used for that purpose. It must also be remarked, that the penetration of the rocket is nearly double that of any ammunition now used in field service, which makes it peculiarly applicable to the attack of all field fortifications, lines, redoubts, &c.; even against 12-pounder field artillery, we should consider ourselves safe behind a sod work of 12 feet in thickness; whereas the average penetration of the 12-pounder shell rocket, at from 800 to 1200 yards distance, is not less than from 20 feet to 22 feet in a solid bank, as proved by the annexed Extract of a Report from Lieut. Col. Fyers; so that the present system of field works would be rendered nearly useless under the fire of such rockets. In short, that grand power of projecting ammunition from the heaviest to the lightest calibres, without apparatus, which gives the rocket its peculiar and characteristic force for application to land service, enables us with any given means to bring either heavier ammunition than any at present used, or a tenfold greater quantity of it by the same means. The details of all these applications will be found in the following pages.

Extract of a Report from Lieutenant-Colonel FYERS.

"A party of Sappers and Miners were employed by Lieutenant-Colonel Jones, Commanding Royal Engineer at Woolwich, to ascertain the penetration of the 12-pounder Rockets * fired into the butt at this place, from a distance of 1260 yards: when they were dug out in his presence as well as mine, it was found that several had penetrated 21 and 22 feet, the shells of which had burst at this depth.

"Troops are supposed to be protected against the fire of field artillery, when placed behind earthen works of 12 *feet thick!*"

(Signed) P. FYERS,
Lieut-Col. Royal Horse Artillery,
Commanding Rocket Corps.

* The Rockets are denominated from their *actual* weight, the same as with the cannon shot.

TABLE OF CONTENTS OF THE TREATISE.

TREATISE,

&c. &c.

THE Rocket is an arm in which there is unquestionably a *combination* of power with *portability*, and a consequent *facility* of application not *existing* in any *other* weapon. As the Rocket contains its own motive force within itself, it may be considered as fixed ammunition, which requires no gun to project it. It may, in fact, in many situations, be discharged *without apparatus of any description whatever;* and where apparatus is required, it is of the lightest sort. Hence, *the Rocket* will be found to *combine all the power of artillery with the facility of musketry:* and thus infantry may be armed with 3 and 6-pounder Rockets, without any burden but the Rocket itself; each man carrying six 3-pounders, or three 6-pounders, and yet having no greater weight to carry than if armed with a musket and sixty rounds of ball-cartridge; notwithstanding *this facility*, however, each of these Rockets will *range as far*, and has *even greater force and penetration*, at 700 or 800 yards than a round of *artillery ammunition* of *corresponding calibre!!!* An infantry regiment of 1000 men, therefore, so equipped, would, *with no greater burden than if armed with muskets*, manœuvre in action with 6000 rounds of 3-pounder, or 3000 rounds of 6-pounder ammunition, which, to all intents and purposes, as to *range, penetration, and effect*, may be reckoned as so much *artillery ammunition* of the same calibre; whereas, to bring into action this same ammunition, in the ordinary way of artillery, in its most formidable* shape, instead of merely involving a common infantry movement, would require the enormous incumbrance of a train of not less than 100 pieces of artillery.

The Rocket combines the power of artillery with the facility of musketry.

Extraordinary comparative facility of transport most important for Indian warfare.

* Of course, the most formidable mode of carrying ammunition is on the gun carriage, which is an *efficient implement* in action, whereas the waggon is a mere *inert machine.*

Power of fire in action equal to facility of transport.

Neither is the *facility* of the *use* of this weapon, and its *consequent power in action*, less extraordinary than its *facility of transport;* for the quantity of Rockets that may be discharged in a flight, *without any apparatus whatever*, being merely laid in rows on the ground, is, in fact, limited only by the quantity of ammunition carried, and by the option of the commanding officer; so that a regiment thus equipped may, as occasion requires it, pour in with all the ease and rapidity of a discharge of musketry, *volleys* of *hundreds* or even *thousands* of *6-pounder ammunition.* Those who have seen the effects even of ten or a dozen such Rockets, thus fired along the surface of the ground, may easily conceive the tremendously destructive operation of 1000 or 500 Rockets thus discharged *at the same moment;* they tear the ground up before them, with all the force of cannon shot in its recochét, and, for the first three or four hundred yards, never rise higher than a man's head. *In such volleys,* therefore, they must *infallibly sweep down and destroy* every thing before them, and that too from the *quantity*, without any *nicety of aim*, more than laying them on the ground in the general direction of the object. This discharge I have denominated "*the ground volley.*"

Use of Rockets by infantry, without carriages.

The ground volley—a rapid manœuvre for discharging 500 or 1000 six-pounder Rockets in flight.

The manœuvre is very simple and rapid, and is as follows: — The men ordered to fire, whether one or two hundred, or more, advance in front of the regiment about twenty paces, with the ammunition, followed by one man to every three ammunition bearers, with their portfires lighted; the ammunition bearers then lay down and point, each man his Rocket, which the portfire men discharge as quickly as possible. By the time the first volley is thus discharged, the ammunition bearers will have prepared a second volley; and thus, from a regiment of 1000 men, formed two deep, the front rank, by advancing a few paces, and each man laying down a Rocket, *a volley of* 500 *may be almost instantaneously discharged;* or, if each man lays down *two* Rockets, *a volley of* 1000, *certainly* with the *same ease*, and very nearly with the *same rapidity* as a volley of musketry; yet, be it remembered, *each round is, in effect, a round of 3 or 6-pounder cannon shot, Shrapnell shell, or case shot!*

Applicable to large Rockets.

But, in fact, the calibre of ammunition which may thus be brought into action is not limited to the 3 or 6-pounders; for, on the same principle, infantry may be equipped to bring into the field 12 or even 18-pounders, without any apparatus; diminishing of course in quantity as the calibre is increased: nay, even the heaviest carcass Rocket for bombardment may thus be conveyed and used, each man carrying a 32-pounder carcass Rocket; so that, by a *very ordinary* besieging army, *ten thousand carcasses,* each, in point of combustible matter, equal to a 10-inch spherical carcass, might be thrown into any town in the course of *one night without trenches, battery, mortars,* or *any of the delays* of a regular siege; discharging, in those few hours, as much conflagrating matter as could be thrown by *ten thousand rounds of 10-inch mortars,* in a *protracted siege of a month,* and, of course, with a proportionably increased effect, as the time is thus reduced. Yet, in all this most powerful mode of bombardment, there is no burden of apparatus to be carried; no mortars, mortar beds, platforms, &c. One or two slopes in different parts, or, what is still more simple, and equally efficient, a few holes, bored by the common miner's borer in the ground at the required angle, to receive the Rocket stick, and thus to point the Rocket for firing; and yet with all this simplicity of application, the extreme range of these bombarding Rockets is not less than 3600 yards. In bombardment.

With the ordinary means of bombardment, taking into account the necessary preparations for bringing up the stores, establishing the batteries, &c. &c. ten thousand rounds of 10-inch mortars could not be brought up and fired in less than a month.

Here, therefore, let us for a moment pause to consider the *enormous differences in the transport,* as well as in the *power of this weapon,* as compared with every other known means; and let us, more especially, direct our view to it with reference to its immense importance in Indian warfare, where the difficulty of the transport of artillery, especially of heavy artillery, is so great! In fact, the importance of this facility of transport holds in all difficult and mountainous countries—and in all expeditions, not only where long marches are necessary, but where a landing on an enemy's coast is involved. Great importance of this arm in India.

But, to proceed in our details. The next use of the Rocket which I shall consider, is its application to cavalry; and here it will be found to possess some advantages which do not obtain when used by infantry, inasmuch as, with cavalry, the new weapon combines more completely with the old. Thus cavalry, armed with Rockets, are still absolutely capable of *all* their other duties as cavalry; the power Use of Rockets by cavalry without carriages.

of horse artillery being *absolutely superadded*, without any deduction of other means.

Equipment of a regiment of cavalry.

In this distribution of the weapon, each trooper carries six 6-pounder Rockets in holsters, and every third man a light tube, with legs, for discharging the Rockets in particular situations, where the ground from being rocky or broken, may not be fit for the ground volley. The manœuvre is very similar to that practised by infantry. The number ordered to fire dismount and advance—and where the ground is suitable, merely lay the Rockets on the surface. Thus a cavalry regiment of 1000 men would carry 6000 rounds of *six-pounder ammunition, and a battery of 330 Rocket tubes**, or *bouches-à-feu;* and would, nevertheless, be as capable of the power of charging in a body, or of pursuing an enemy, as at present.

Rapid use of the tubes.

[Where the tube is used, it is quite immaterial what may be the nature of the ground fired over, whether ploughed ground or rocky, or across a ravine or river. The Rocket from this tube ranges clear of the surface, to distances regulated as in artillery by the elevation: nor have I any hesitation in saying, that taking a series of rounds of field artillery and Rockets *fired in action*, an equal if not a greater number of Rockets would take effect; because the Rocket tube, *once properly laid*, requires no *fresh aim*, there being no recoil or re-action in firing a Rocket,] while the gun must *be pointed after every round;* an operation frequently *quite impracticable* in the smoke and confusion of battle, and when it *can* be done, *always attended with considerable delay*. At a review, indeed, artillery appears capable of very rapid firing; but if it were often to be so practised in battle, very few rounds would take effect. The operation of loading a Rocket tube is certainly more simple than that of loading a gun; the Rocket tube might therefore unquestionably be fired as quick if not quicker than the gun, merely by regulating the supply of ammunition accordingly: in doing which, from the extraordinary facility, of its transport, it is evident that there need never be any deficiency!

Do not require fresh aim in action, having no recoil.

Comparison between the powers of the Rocket system and those of artillery.

And here again I cannot resist drawing a comparison of the power which may thus, by the Rocket, be *superadded* to *any* cavalry regiment, as contrasted with the powers of artillery.

330 six-pounder bouches-à-

We have seen, that a cavalry regiment of 1000 troop horses, *without a single wheel-carriage*, carries into action, and will manœuvre in the

* The *weight* of one of these tubes for discharging a *six-pounder Rocket* is less than that of a *musket*.

same space as any other cavalry regiment, 6000 rounds of *six-pounder ammunition*, and a battery of 330 *bouches-à-feu*, *without sacrificing any of its powers as cavalry.*

feu may be brought to bear upon *one point* by a regiment of cavalry.

Now, *merely* to *move* a battery of 330 six-pounder guns would require 1320 horses; but as to using them for any *concentrated operation*, such as that of which the *battery* of a cavalry regiment *armed with Rockets* is capable, *it is utterly impossible!* Three hundred pieces of artillery, drawn up in line, would occupy nearly three miles; so that any point assailable by the centre of this battery would be beyond the range of either flank; indeed it is a most difficult matter to bring even fifty or sixty pieces of artillery to bear upon any point, and can only be effected by a slow and preconcerted operation; but as to any rapid or concentrated movement of artillery, with a regiment of cavalry, such is the space required for such movements, that more than *six* pieces of ordnance could not be so combined. Yet a regiment of cavalry, *armed with Rockets,* has the power of bringing into action this battery of 330 six-pounder *bouches-à-feu,* with the greatest ease and facility, in the concentrated space which it occupies when halted, and in all situations, let the ground be what it may; while, in most situations, where cavalry can act, this battery may be increased to *500 or 1000 rounds in a volley.* Such discharges are evidently not required every minute, though several of them might be so repeated if necessary. Surely, therefore, it is not too much to say that one or two regiments, having this power, might decide the fate of any battle! No steps, however, have as yet been taken towards the establishment of either of these modes of using the *Rocket,* though its *most simple* as well as its *most efficient* application.

Utter impossibility of concentrating the same number of bouches-à-feu by artillery.

Hitherto we have supposed the Rocket used, merely by infantry and cavalry, without the aid of any wheel carriage whatever; now we have to show *its application by wheel carriages:* and here again, although the *peculiar* and *most formidable characteristic* of this weapon is its applicability *without carriages,* still it will be found, that where *carriages are used,* they also afford, in their way, powers peculiarly appertaining to the Rocket system.

Use of Rockets by wheel carriages.

In determining the principle upon which these two opposite modes should be applied, the following rule may be laid down; — viz. that where *large bodies of men* are to be armed with Rockets, such as a regiment either of infantry or cavalry, they should be equipped as

General rules for the distribution of this arm in different cases.

above detailed *without carriages;* where *small bodies* of men are to be armed, such as a troop of cavalry for instance, they should not only carry a proportion of Rockets on horseback, but be equipped also with *carriages.* The regiment is adequate, without any *extra means,* to carry an *immense abundance of ammunition,* while, by being unincumbered with carriages, its general powers are no ways crippled:—on the other hand, the smaller body, by having carriages, attains the power of conveying also an *immense supply* of ammunition, *in proportion to its numbers,* and of bringing into action a tremendous battery, which it could not otherwise possess. The regiment, in fact, obtains these powers sufficiently *without carriages,* and preserves all its force as cavalry; the troop obtains its utmost power as artillery *by the carriages,* and has no power as cavalry to sacrifice; hence, therefore, the rule I have laid down.

Principle on which the *small* corps hitherto equipped have been armed to *make the most of them.*

It is on this principle, therefore, that the mounted Rocket troop, the only force as yet appropriated to the service, has been armed. The number of men allotted to the service having been *small,* I have been induced to recommend that mode of equipment which would give them the *greatest possible* force, *in proportion to their numbers;* and here, also, though I am far from considering this application as being in the *true spirit* of the Rocket system, it will be found to give most extraordinary powers to a handful of men.

Formation of a Rocket troop.

Thus it will be seen, by reference to page *55,* that a *troop,* on the *usual establishment of horse artillery,* consisting of 97 gunners, 36 drivers, with the 6 Rocket carriages, and 6 ammunition waggons, will convey 4120 rounds of ammunition, and 142 bouches-à-feu; whereas a troop of horse artillery would carry into action, *with the same means,* only 1002 rounds of ammunition, and *six* bouches-à-feu; so that the ammunition conveyed by *one* Rocket troop is equal to that conveyed by *four troops of horse artillery:* and the *number* of bouches-à-feu, and *power* of throwing in a *great quantity of instantaneous fire* upon any point, *far exceeds* that which *ten* troops of *horse artillery* could produce. Yet, taking the *average* of the calibre of the Rocket ammunition, consisting of 18, 12, 6-pounders, &c. it will be found equal to that of the ammunition carried by a troop of horse artillery.

Self-motive power of the Rocket, with-

Now this extraordinary difference of power, and consequent economy of means, arises in this, as in *every other* application of the Rocket, from its *self-motive property*, and from the consequent *lightness of the tube* used to project it, as compared with the weight of the field gun required to project a corresponding round of artillery.

out ordnance, the cause of all its portability, power, and economy.

The weight of a 12-pounder gun is 18 cwt. while that of a 12-pounder Rocket tube, which projects the *same weight* of *ammunition*, and *at least* to the *same distance**, is only 20lbs.; so that, with the *same number of horses*, *seventy* bouches-à-feu of the Rocket 12-pounders might be carried, for *one* of the 12-pounder guns.

Multiplication of 12-pounder bouches-à-feu, by the Rocket system, as compared to artillery, 70 to 1, with the same number of horses — with the 9-pounder, 90 to 1, and with the 6-pounder, 50 to 1.

In like manner, the weight of a 9-pounder gun being 13 cwt., while that of a 9-pounder Rocket tube is only 16lbs., *ninety* to *one* might be carried into action by *the same means*. So also, the weight of a light 6-pounder gun being 6 cwt., and that of the 6-pounder Rocket tube only 13½lbs., *fifty* to *one* might be carried into action by the same number of horses.

In constructing, however, the field carriages for the use of Rockets, I have taken advantage of this difference of weight, to carry *not only a greater number of bouches-à-feu* with a given number of horses, but *a greater proportion of ammunition also*: conceiving it more expedient for general purposes thus to augment the power of *supply*, as well as that of *discharge*, rather than to lay all the increase upon the latter power only.

According to this rule, each 12-pounder Rocket carriage may be constructed to convey 10 bouches-à-feu, and 160 rounds of 12-pounder Rockets, with *the same carriage and number of horses* that would carry only *one* 12-pounder gun, and *twelve* rounds of 12-pounder ammunition. But as the Rocket carriage, from experiencing no recoil in the discharge of the Rocket, may be made much lighter, a 12-pounder Rocket carriage, expressly constructed for this purpose, may be made to convey a battery of *twenty* bouches-à-feu, and *one hundred and sixty* rounds of ammunition, with no greater weight for the horses than a 12-pounder gun with 12 rounds.†

A 12-pounder Rocket carriage conveys *twenty* bouches-à-feu and 160 rounds — a 12-pounder gun carriage conveys only *one* bouche-à-feu and *twelve* rounds.

Thus also a 9-pounder Rocket carriage may be made to carry 20 bouches-à-feu instead of *one*, and *one hundred and seventy* rounds instead of *thirty-two*.

And a 6-pounder Rocket carriage may be made to carry 20 bouches-à-feu instead of *one*, with 112 rounds of ammunition instead of *thirty*, with four horses.

* From 1200 or 1500 yards, at low angles, and 2500, its extreme range.

† The 12 and 9-pounder gun carriages are each drawn by six horses.

The *volley carriage* travels ready charged, and discharges 20 Rockets the moment it is unlimbered

The Rocket carriages so constructed are called *Volley Carriages*. The tubes of these carriages *always travel charged;* so that, at the moment of unlimbering, each carriage is prepared to pour in a volley of 20 Rockets. These carriages have not only been used in this country, but I have constructed pattern carriages, which have been transmitted to India.

This, however, is not the only advantageous modification of the Rocket carriage, depending upon these peculiar properties of the weapon. By a different arrangement it may be applied to *economise* the *number* of *horses*, which, in some cases, may be more important than carrying the power of discharge and supply to its utmost extent, and retaining the *same* animal force that is required for the carriage of artillery of corresponding calibre.

Great additional powers may be retained with half the number of horses.

Thus a 12-pounder Rocket carriage may be constructed to travel with *two* horses instead of *four*, and yet will carry into action 2 bouches-à-feu and 40 rounds of ammunition; a 6-pounder, also with 4 bouches-à-feu and 60 rounds, and a 3-pounder with 8 bouches-à-feu and 120 rounds of ammunition, may be drawn by a pair of horses each. And I have, indeed, constructed 3-pounder, 6-pounder, and even 12-pounder Rocket carriages, which may be *carried into action* and manœuvred *entirely by men*, without any horses at all.

The true principle of the economy of this weapon is, to give it to existing corps without raising troops expressly for it.

Nevertheless, I must here repeat, that notwithstanding the *extraordinary powers and facilities* of carrying *heavy ammunition* into action, thus given to a *mere handful of men*, still I consider the *true* application of the weapon to be that which places it in the hands of the *already existing corps* of every army, instead of embodying corps expressly for it. It is thus that we obtain its *most important property of universal application*, without the *diminution* of *any other existing powers*, and consequently without *extra expense*, either in the additional pay of troops, in horses, or carriages; the first cost of the ammunition, and that of the few trifling * articles of equipment, such as tubes, holsters, &c. being, in fact, the only expense.

The duties of the small corps, formed expressly for the Rocket service, should be principally limited to the more scientific operations for the attack of forts and field-works, where the heavier shell Rockets

* A *six-pounder portable* tube, *complete*, costs from 2*l*. to 3*l*.— a six-pounder gun costs 72*l*. *without* the carriage, horses, &c. &c. to convey it.

are to be used, and where the objects of attack being smaller, greater nicety is required.

Before I conclude this part of my subject, I must state, that various other descriptions of Rockets, which have not yet been mentioned, have now been made, from the weight of *two* ounces up to three hundred pounds. The *two-ounce* Rockets are a species of self-motive musket-ball cartridges, which require a musket of *only four* pounds' weight, though the missile itself is twice the weight of a musket-ball, yet having its range and penetration at least equal. Hence from the *lightness* of the musket, though the ammunition is so much *heavier*, a soldier, with *ninety* rounds of two-ounce Rockets, and the appropriate musket, has no greater weight to carry than with the common musket and *sixty* rounds of ball-cartridge! On this same principle small Rockets have been made from two ounces up to *half a pound*, to be discharged *from the shoulder ;* and of *one pound* weight, to be projected from light tubes, formed as lances, either for infantry or cavalry ; so that a lancer, without interfering with any of his ordinary duties, would thus have the power of projecting a missile at his enemy equal to the round of a one-pounder piece of artillery, called the Amusette. The one-pound Rocket possesses all the range, weight, and force of the one-pound shot, so that a regiment of lancers, each man being so armed, would possess a tremendous fire, and carry 15,000 or 20,000 rounds of this ammunition into action without incumbrance!!!

Rockets have been made from two ounces up to three hundred weight.

Advantages of the small Rockets compared to musket cartridges.

It is evident, therefore, that the weapon still possesses its peculiar powers, even in its most diminutive form ; while as to the other extreme I am convinced, that *three hundred weight* is by no means the utmost mass that may be projected by the force of the Rocket, although *it very considerably exceeds* in weight any missile now discharged by gunpowder. The limit as to the weights that can be projected by the ordinary use of gunpowder is not only prescribed by the *difficulty* of *casting* the *discharging instrument*, the gun or the mortar, but that of *moving it* after it has been cast; thus the 13-inch shell, which weighs only *two hundred pounds*, requires a mortar of *five tons* to project it. As to the Rocket, however, neither of these limits exists, but no such instrument as either gun or mortar is required for land service; inasmuch as the Rocket of *three hundred weight* may be thrown *without any apparatus* at all, from an earth-slope; and for sea-service, that is to say, for projecting these large Rockets from ships or boats*, where apparatus would be required, the weight of it may be considerably *less* than that *of the Rocket itself.* Even, therefore, if Rockets of still greater dimensions than three hundred weight were made, which, indeed, I contemplate, amongst the many extensions yet practicable, for the purpose of breaching walls, by impelling against them masses of ten hundred weight, or *even more—these masses, impossible as it is* ever to *use them* as missiles on the *ordinary principles* of artillery, may evidently be employed, by our armies, without any serious incumbrance.

Thus, therefore, I may state, that, to whatever extreme this weapon is carried, I still find the *great* and *marked peculiarities* of property, on which its power and facility depend, predominant!

Concluding remarks as to the great powers of the Rocket system.

I feel myself, therefore, justified in closing this part of my subject by observing, that that decisive system of warfare, which it has been the aim of all great generals to accomplish, and by which it is attempted to bear down and annihilate every thing at the *devoted* point, is the *inherent* and *peculiar property* of the *Rocket system;* and that *effects* both as to the *quantity of ammunition* conveyed and discharged, and as to its *weight, range,* and *penetration,* may be produced by this arm, with *all the ease and facility* of the *common manœuvres* of infantry and cavalry, *which it is utterly impossible at all to approximate to,* by *any* application of artillery *whatsoever:* yet I feel that I may defy any fair and candid person to assert that I have overcharged a single statement; if any one, indeed, is inclined to say so, I promise him I will not shrink from the proof, nor turn my back on any discussion, in which the writer will avow himself, though I shall ever treat such anonymous assertions, as have hitherto only appeared against this system, with the contempt due to that disguise, which is ever assumed as the shield either of ignorance or falsehood!

Consideration of the great economy of the system.

So much, therefore, for the *powers* of this weapon! I shall now proceed to some few details as to its *economy.*

Mistaken notion as to the expense of

It has been said, that the Rocket is a most expensive weapon, and that its first cost was enormous: but if the *difference of the cost* of the Rocket ammunition, compared with that of artillery, were *ten times*

* It is another of the extraordinary faculties of this weapon, that from having no recoil or re-action in its discharge, the smallest boats are capable of throwing the largest Rockets.

what *it is*, the absence of the enormously expensive charges attendant upon artillery in *its transport*, as well as in the first cost of its *requisite apparatus*, and in the *wear* and *tear* thereof, *would be found to yield an immense saving*. *The fact*, however, *is*, that the *cost of the Rocket* itself is, in some descriptions of ammunition, and in the heaviest calibres, *even less* than the corresponding rounds of artillery ammunition.

Rockets confuted.

First cost of Rocket ammunition, in some cases, *less* than that of *artillery ammunition*.

Thus the cost of a 13-inch spherical carcass, with the charge of powder required to throw it 3000 yards, is £2 1*s*. 1*d*. without reckoning any thing for mortars, mortar-bed, platform, &c.; while that of a 42-pounder Rocket carcass, which requires *no apparatus whatever* to project the same quantity of combustible matter to the same distance, is only £1 19*s*.

So also, in comparing the shell Rocket with the Shrapnel shell, round for round, the cost of the larger natures of the Shrapnel shell exceeds that of the corresponding natures of shell Rocket, and is but little less in the smaller natures.

Now, when it is recollected that the Shrapnel shell requires all the *expensive apparatus of artillery* to project it; and that, on the other hand, the shell Rocket may, as we have seen, be used in quantities, at least *one hundredfold greater* than the Shrapnel, either by *existing* infantry or cavalry regiments, and *absolutely without any additional charge for its use*, or *without deducting* from the *other duties* of the men employed — I trust that little doubt can remain of the *economy* of the Rocket system; not only, indeed, where the comparison is made with the Shrapnel shell, in which the difference of first cost *is so trifling*, but even where common round shot are used, in which, certainly, there is a considerably greater difference in the relative first prices: yet will the difference in this case also vanish, when *the balance of the other expenses* incident to the *actual* use of the *cheapest* artillery ammunition is brought into the calculation. It is an absolute fact, that, during the Egyptian campaign, every round of ordinary artillery ammunition, on an average, cost the government £20!

Even in the most expensive use of the Rocket, that of the establishment of *the troop*, armed with carriages, &c. constructed on purpose, such, we have seen, is the *comparative cheapness* of the *apparatus*; such the difference of expense in the *number* of *horses* required to carry

Heavy expense *inseparable* from the *use* of artillery.

into the field the *same number of bouches-à-feu*, and the *same quantity of ammunition;* that no length of transport, no local circumstances whatever, could raise the charges of the use of the Rocket *in any degree at all* to be compared to those which are *inseparable* from the employment of artillery.

The Rocket, if properly employed, involves little or no expense beyond the first cost of ammunition.

But, I repeat, if the Rocket be *used* on its *true principle*, that is to say, by distributing it largely *to all existing corps* — if, as I hope soon to see practised, the flank companies of the infantry regiments, and two squadrons of the cavalry regiments, shall be armed with Rockets, so as, in fact, to give the whole army the *power* of *artillery every where* and the *expense* of it *nowhere* — then, indeed, there will be *no extra charge* to add to the *first cost* of the ammunition, and the *full benefit* of the *economy* of the system will be derived.

The Rocket is, in truth, an arm by which the whole system of military tactics is destined to be changed. The power of using *the heaviest projectiles of artillery, with the facility of musketry*, yet without any diminution of either *force* or *range*, and with all the precision requisite for general action, is a *power* which is not to be withstood, as armies are at present constituted and manœuvred. The great military states of Europe are now striving to obtain this weapon, and ultimately of course they must succeed.

Answer to the objection, that it is dangerous to *ourselves* to encourage the use of this weapon.

This, I am aware, has been stated as a source of objection to the invention. In answer to it, therefore, I shall here adduce the following arguments: — 1st. As the only limit to the quantity of fire that can be brought to bear on any point, is the quantity of Rockets that may be *accumulated* at that point, it is evident that the Rocket, however *powerful in attack*, is still more so *in defence* — because there really is *no limit* to the quantity of ammunition that may thus be *accumulated*, to be poured forth for the *defence* of any *important post or pass*. This property, therefore, certainly makes the Rocket, if properly arranged, an arm more to be *desired* than *feared*, in all situations. Another property of the Rocket having a similar tendency is, that it makes the few, who have the pecuniary means, equal to the many. Even, therefore, if this property went to establish an equilibrium between the power of large and small states, England, with her *limited* population, but possessing at the same time unlimited means of thus accumulating the *mechanical* powers of war, would be rather a gainer than a loser by it — while it is evident, also, that the general cause of humanity must gain by every military invention which neutralizes the efforts of *mere numerical* superiority. As good order in society is promoted by the refinements of the modern practice of duelling, so, on the great scale of nations, peace is preserved by those improvements in the art of war which render brute force less availing.

But, even if it were a well-founded objection, it is now too late to make it. The powers of the Rocket are too well known and acknowledged to allow it ever to sink into oblivion, and our obvious policy is, therefore, to keep the lead which we have attained in the possession of this weapon, not only by urging every possible improvement, but by extending its organization and use to the utmost.

I shall conclude this sketch by adducing some very important proofs of the durability of the Rocket Ammunition.

Proofs of the durability of the Rocket.

In the first place, then, I must refer to an experiment made some few years since to ascertain the state of preservation of some of the first constructed Rockets; part of which had been made *upwards* of *seventeen years!!* had been sent on various expeditions to different parts of the world — to the Baltic, to North America, to the River Plate, and to the Mediterranean, where they had remained for several years in store. Greater vicissitudes of climate, or longer voyages, could not well be found; and yet these Rockets, at the expiration *of seventeen years*, the last five of which they had been kept in a casemate in Dover Castle (no very dry magazine), were found by every possible mode of proof to have undergone no perceptible deterioration since their first making; insomuch that the Board of Ordnance determined on still keeping them in store for *future service*. What, let me ask, would have been the state of any other known ammunition under similar circumstances? The truth is, that the Rocket being enclosed in a metallic case, impervious to air or moisture, with common care, is scarcely susceptible of injury, and I have accordingly fired Rockets after they have been under water for several hours.

Experiment showing the perfect preservation of Rockets after being made 17 years.

When such proofs are produced of the general durability of this weapon, it can, I think, scarcely be supposed, that, with common and ordinary precaution, even a voyage to India can deteriorate them;

and, when once they are delivered in good condition, the climate will have, of course, no more effect upon them after the voyage than if they were made in that country.

But as these may be considered to be rather proofs by induction than direct, I shall proceed to show, by the actual state of three equipments of Rockets, which have been sent to the Cape of Good Hope, to Madras and to Bengal, that no length of voyage or climate affects them. In the first place, therefore, I shall insert the following Report from Major Dundas, which was transmitted to me by the Governor, Lord Charles Somerset.

"*Cape Castle, Aug.* 12. 1818.

"SIR,

Report of the condition of Rockets, after being kept five years at the Cape of Good Hope.

"Not having received a Report of the trial of Rockets from Graham's Town until last post, I have been unable, hitherto, to make the general Report, for which I have been called upon by His Excellency's (the Governor's) commands. I have the honour now to state for his information, that the Rockets, in store, here and at Graham's Town, are in a perfect state of preservation, and that they do not appear to have suffered in the slightest degree from the climate. Of eighteen of the different kinds which have been submitted to trial, only *one* proved defective, but this I attribute to some accidental faultiness in its original formation.

"The ranges of the larger, or twelve-pound Rockets, varied very little, and may be stated at 7000 feet; those of the cavalry, or three pound Rockets, at 4000; which, compared, with the ranges of new ones tried at Woolwich, prove that no sensible injury has been sustained by them from age or any other cause.

"I have the honour to be, SIR,

"Your most obedient and humble Servant,

(Signed) "W. B. DUNDAS,

"Major, Commanding Royal Artillery.

"To Major ROGERS, &c. &c. &c."

We now come to the consideration of the effect of climate upon Rockets. The following is a Report of practice with some of the improved Rockets, made at the manufactory at Bow, and transmitted to Calcutta, in the Timandra, in 1820.

Report of Rocket practice at Meerut, showing the perfect state of Rockets after being three years in store in Bengal.

Return of Rocket Practice carried on at Meerut, before Major Gen. T. Reynell, C.B. Commanding, on the 29th May, 1823.

Nature of Rockets.	No. of Rounds.	No. of Boxes whence taken.	Distance of Object.	Object.	Elevation.	Rockets whence fired.	State or Appearance of Rockets.	Nature of Shafts.
					8			
3-Pounder	1	108	500 Yards.	4 Flags, at 20 Yards distance from each other, presenting a Front of 60 Yards.	6	Six-Feet Tube	In good order	Bamboo.
	2	...			2	Ground Chamber	Ditto	Ditto.
	3	...			6	Tube	Ditto	Ditto.
	4	63			2	Chamber	Very good order...........................	Ditto.
	5	...			6	Tube	Good	Ditto.
	6	...			2	Chamber	Very good	Ditto.
	7	12			6	Tube	Good	Ditto.
	8	...			2	Chamber	Slightly mildewed...........................	Ditto.
	9	...			6	Tube	Very good	Ditto.
	10	105			2	Chamber	Slightly mildewed...........................	Ditto.
	11	...			6	Tube	Vent a little rusty	Ditto.
	12	...			2	Chamber	Good	Ditto.
	13	90			6	Tube	Much mildewed.	Ditto.
	14	...			2	Chamber	Good	Ditto.
	15	...			6	Tube	Slightly mildewed, and vent a little rusty .	Ditto.
	16	70			2	Chamber	As No. 13...........................	Ditto.
	17	...			6	Tube	Very little mildewed...........................	Ditto.
	18	...			2	Chamber	Much mildewed...........................	Ditto.
6-Pounder	1	257	700 Yards.		45	Light Infantry Frame	Good	Deal.
	2	...			7	Six-Feet Tube	Ditto	Bamboo.
	3	...			2	Ground Chamber	Ditto	Ditto.
	4	180			7	Tube	Slightly mildewed...........................	Ditto.
	5	...			2	Chamber	Vent a little rusty...........................	Ditto.
	6	...			30	Light Infantry Frame	Much mildewed, and vent much rusted ...	Deal.
	7	272			45	Ditto	Vent slightly rusted	Ditto.
	8	...			7	Tube	Ditto	Bamboo.
	9	...			2	Chamber	Much rust and mildew...........................	Ditto.
	10	271			7	Tube	Very little mildewed...........................	Ditto.
	11	...			2	Chamber	More mildewed...........................	Ditto.
	12	...			30	Light Infantry Frame	Very much mildewed.	Deal.
	13	230			45	Ditto	Very good	Ditto.
	14	...			7	Tube	Ditto	Bamboo.
	15	...			2	Chamber	Ditto	Ditto.
	16	252			7	Tube	Ditto	Ditto.
	17	...			2	Chamber	Ditto	Ditto.
	18	...			30	Light Infantry Frame	A little oily	Deal.

Remarks. — Those fired from the Light Infantry Frames were not directed at any Object; all the others, excepting perhaps three or four, passed within the Flags, and generally at an efficacious Height. Those from the Ground Chambers, it was remarked, ranged truest, and with the best Effect. The Result of this Day's practice was most satisfactory, not a single Rocket having burst, after sustaining a *second* rainy Season. Boxes Nos. 105 and 90, containing 3-Pounder Rockets, were purposely taken from the most exposed Situation in the Magazine, to see whether they had been affected by the Humidity of the Atmosphere at that season. — No Attempt was made to remove Mildew, but the Vents were cleared of Rust. — The whole were fired by a common Portfire without Quickmatch.

(Signed) C. GRAHAM, Captain,
Commanding Seventh or Rocket Troop Horse Brigade.

True Copy — H. NICHELSON, Captain,
Superintending New Rocket System.

By this Report it will be seen, that out of thirty-six Rockets fired on that occasion, *not one of them failed;* this practice was carried on at Meerut, in the presence of General Reynell, on the 29th of May 1823.

These were some of the first of the improved Rockets transmitted to India. After the voyage to Calcutta, they were sent up the Ganges some hundred miles, to Meerut, the head-quarters of the Rocket troop, where they had been kept without any particular care, as was evident from the appearance of them noted in the Report.

Similar practice, with similar results, was carried on at the Mount, Fort St. George, in the presence of the Commander-in-Chief, Sir A. Campbell, of an equipment of Rockets, made by me, and sent out to Madras, in 1822; in consequence of the success of which, it appears that the Rocket troop, which had been dismantled in 1821 for want of ammunition, has been since re-established in its original scale. The following letters, relative to this practice, need no comment.

Copy of a Letter from His Excellency General Sir ALEXANDER CAMPBELL, Commander-in-Chief of His Majesty's Forces at Madras, to Captain H. NICHELSON, Inspector of the New Rocket System in India.

Madras, Dec. 7. 1822.

"MY DEAR SIR,

Documents from Madras, showing the perfect state in which the last Rockets arrived there in 1822.

"It is difficult to express to you the gratification I received from the exhibition of the Rocket practice yesterday evening at Saint Thomas' Mount.

"The great improvements effected by Sir WILLIAM CONGREVE, in the composition and construction of this powerful and destructive implement of war, has entirely removed the strong prejudice that existed (until your arrival here) against it in this army, from the inferior nature of the Rockets used during the last war. I shall take an early opportunity of recommending to this government the re-establishment of a Rocket troop; and I feel that this army is much indebted to the Marquess of Hastings, for his great consideration in sending to instruct us an officer so highly qualified as you are, in the use of this important weapon.

"I have the honour to be,
"With the greatest esteem and regard,
"Your faithful humble servant,

(Signed) "A. CAMPBELL.

"A true Copy.
(Signed) "H. NICHELSON."

Extract of a Letter from Captain NICHELSON, Inspector of Rockets, to Sir W. CONGREVE.

"After having given over all expectation of the Providence, she made her appearance in the roads on the 17th of last month. As soon as the Rocket apparatus could be got on shore and sent to the Mount, I commenced operations, which I am happy to say I have carried on most satisfactorily, up to this day, *without one Rocket bursting* or in any way unfavourable. *The artillery are delighted with them.* I have given Colonel J. NOBLE, commanding the horse-artillery, a copy of the formation of the Woolwich and Bengal Rocket troop. I have no exact Report of practice that I can send at present, but *more than half* of all that were fired *went through the curtain.* One of the 12-pounders penetrated a tree of more than two feet diameter, burst in it with capital effect, which proved to the numerous spectators the great power of the weapon."

A few words only are now required as to the sufficient accuracy of the weapon for all service in the field. This fact is best proved by the Extracts from the Documents inserted: — thus Capt. Nichelson informs us, that more than half the Rockets fired at the Practice at Mount St. Thomas went *through the curtain.*

I cannot, therefore, but believe, that a candid consideration of the statements and documents contained in this TREATISE must fully prove the great powers and facilities of the weapon, its extraordinary penetration and range, and the accuracy to which it is now brought. At the same time that it is shown to be the most economical and the most durable description of ammunition that can be used.

GENERAL INSTRUCTIONS FOR THE USE OF ROCKETS,

BOTH

IN THE FIELD AND IN BOMBARDMENT,

SHOWING THE SPIRIT OF THE SYSTEM, AND ITS COMPARATIVE POWERS AND FACILITIES.

It must be laid down as a maxim, that "the very essence and spirit of the "Rocket System is the facility of firing a great number of rounds in a short "time, or even instantaneously, with small means," arising from this circumstance, that the Rocket is a species of fixed ammunition which does not require ordnance to project it; and which, where apparatus is required, admits of that apparatus being of the most simple and portable kind.

An officer, therefore, having the use of this weapon under his direction, must ever bear this maxim in mind — and his first consideration must be — to make his discharges against the enemy in as powerful vollies as he possibly can.

Thus, if the defence of a post be entrusted to him, and the ground be at all favourable, he will, independent of the regular apparatus he may have at his disposal, prepare what may be called Rocket Batteries, consisting of as many embrasures as his ground will admit; these embrasures being formed by turning up the sod, so as to give channels of direction four or five feet long, and three feet apart, by which a great number of Rockets in a volley may evidently be arranged to defend any assailable point. In these embrasures, if liable to surprise, the Rockets may be placed in readiness, the vents *not* uncovered; though, generally speaking, this is not necessary, as so short a time is required to place them, — here and there one, only being in its embrasure.

In battle, also, where there is not, of course, time to prepare the ground as above stated, but where it is tolerably level, he may, in addition to the apparatus he possesses, add to his fire by discharging, from the intervals of his frames or carriages, Rockets merely laid on the ground in the direction required; and, if an enemy be advancing upon him, there is, in fact, no limit to the volley he may be prepared thus to give, when at a proper distance, but the quantity of ammunition he possesses, the extension of his own ground, and the importance of the object to be fired at. Under these limits, he may choose his volley from 50 to 500 — a fire which, if judiciously laid in, must nearly annihilate his enemy: for this purpose trains are provided. This practice also requires the exposure of only one or two men, who are to fire the volley, as the remainder, with the ammunition, may be under cover. And here it should be remarked, that the length of ranges, and the height of the curve of the recochét, in this mode of firing, depend on the length of the stick — the stick of the full length giving the longest range, but rising the highest from the ground; the reduced stick giving a shorter range, but keeping closer to the ground. From this application, therefore, where practicable, by carrying a certain number of the 12-pounder pouches in the ammunition-waggon, an officer, even with a dismounted brigade, may always manœuvre and detach parties to get upon the flanks of any approaching or fixed column, square, or battalion, while he himself remains with the heavier ammunition and carriages in front.

This mode of firing from the ground of course applies only for moderate distances; the limits of which, with the smaller natures of Rockets, may be considered from 800 to 1000 yards, and for the larger from 1000 to 1200; where, therefore greater ranges are required, the apparatus must be resorted to. And here it is proper to remark, that in the use of the Rocket, at least in the present state of the system, no certain increase of range can be depended upon by increasing the elevations from the ground-ranges up to 15°, for the smaller Rockets; and 20° to 25° for the larger; for in the intermediate angles the Rocket is apt to drop in going off, and graze near the frame; but at the above angles it will always proceed in a single curve to very greatly increased ranges, from 1500 to 2000 yards.

In bombardment, as well as in the field, the quantity of instantaneous fire is equally important, and the greater number of Rockets that can be thrown not only increase the number of fires, but, by distracting the enemy's attention, prevent their extinction. To this end, therefore, an officer should always employ as many bombarding frames as possible; and here again he will find, that in bombardment, as well as in the field, the weapon affords him the means of extending his fire beyond the compass of his apparatus.

Thus, he may form a Rocket Battery, of any common épaulement, parallel to the face of the town to be bombarded, by digging a trench in the rear of it to admit the stick, so as to lay the Rocket and stick against the slope of the épaulement, that slope being brought to the desired elevation for projecting the Rocket, or by boring holes to receive the stick; or he may construct a slope expressly as a Rocket Battery; and as, in firing these vollies, his Rockets need not be more than three feet apart, it follows, that from an épaulement or battery of this description, fifty yards in length, he may keep up this bombardment by a discharge of fifty Rockets in a volley, and repeat these vollies every five minutes if desirable; a rate of firing which must inevitably baffle all attempts of the most active and numerous enemy to prevent its effect.

It is obvious, therefore, that in any comparison made of the powers of the Rocket with those of common artillery, whether an officer be called on merely to demonstrate its powers, or to carry it actually against an enemy, the foregoing maxim must be his rule; in fact, every thing should be demonstrated according to the spirit of its use; a single Rocket is not to be compared with a single gun-shot, by firing it at a target. But the consideration is, whether for general service the power of quantity in the fire of Rockets does not *at least* counterbalance the greater accuracy of the gun? and for this purpose the spirit of the demonstration of the Rocket system is to show how few men are required to produce the most powerful vollies with this arm. No demonstration should be made with less than twenty rounds in a volley; to maintain which, in any fixed position, at the rate of two or even three vollies a minute, twenty men may be said to be sufficient, and this with Rockets projecting cohorn, or $5\frac{1}{2}$-inch howitzer shells, or even 18 and 24-pounder solid shot. The first point of comparison, therefore, is — How many rounds of *such* ammunition in the minute could twenty men project by the ordinary means of artillery?—or how many in a volley, even if they had all the means at hand? — And the next point is — What are the comparative *facilities* in bringing these different means into action, where the one system requires only the transport of the ammunition—the other, not only that of the ammunition, but of the most massive ordnance, without which it is entirely useless?

But independent of this comparison as to quantity, there are others in which the Rocket has advantages exclusively its own: there are situations where artillery cannot by any means be brought into action, while there is no situation, no nature of ground, which is passable to an infantry soldier with his musket in his hand, that is not equally to be passed by the Rocketteer with *his* arm and ammunition. For the accomplishment of any particular service, he may dispense entirely with wheel-carriages or even horses; there is nothing which the men themselves cannot transport and bring into action; and if any bombardment were required by a *coup de main*, 1000 men would not only convey 1000 rounds of the heaviest Carcass Rockets, a number sufficient to destroy any place within the compass of their range, but would perform that service in a few hours, having neither batteries or platforms to erect, nor mortars to convey.

Such are the true principles of this new system of artillery, for (projecting the same ammunition) so it may be called, and the greater the scale of equipment, the greater in proportion will its powers appear; thus, if an establishment were formed on the strength of a cavalry regiment, if 600 mounted men were equipped on the principles of the present detachment, they would take into action, without ammunition-horses or wheel-carriages, 2400 rounds of ammunition, and 200 bouches-à-feu; and if 100 ammunition-horses were attached to this corps, it would further possess a reserve of nearly 2,000 rounds more; the whole capable of every movement and service practicable by any other regiment of heavy cavalry; and the same proportionate power would be found to attach to every other mode of equipment.

In addition to this view of the powers of the weapon, it is important to state, that the detail of the service is most extraordinarily simple; that there are but a few points to be attended to in its application, and those such as may be most easily acquired: the principal ones are, that care must be taken to fix the sticks very firmly to the Rocket, and in the true direction of the axis of the Rocket, to prevent aberration of flight.

That, at high angles, the frame must always be elevated for the large Rockets from 5° to 10° more than the elevation at which the Rocket is intended to be projected, and in the small Rockets from 2½° to 5°; for, as the Rocket leaves the frame before it has obtained its full force, it drops a certain number of degrees in proportion to its weight at going off. Thus the longest ranges of the 32-pounder Carcass Rockets are obtained at about 55°, or rather more, if the Rockets have been long made. An officer, however, being prepared for this circumstance, will soon discover the maximum range of the Rockets he may have to discharge.

Some allowance in elevation also must be made for the direction of the wind; if it is powerful, and blows in a contrary direction to that in which the Rocket is projected, the frame requires *more* elevation; for the wind, acting more on the stick than the body of the Rocket, depresses the elevation in its rising. If, on the contrary, it blows in the direction of the Rocket's flight, *less* elevation is required; for, in this case, the Rocket mounts by the wind's action on the stick. So, from the same cause, if the wind be strong, and across the range, though no difference of elevation is necessary, still an allowance must be made to leeward; for the Rocket, contrary to the course of ordinary projectiles, has a tendency to draw to windward: a few rounds, however, in all these cases, will immediately point out to the observant officer what is the required allowance. These remarks refer only to high angles; for no effect whatever is produced by the wind in the ground-ranges: in these the only caution necessary to be attended to is, to choose the most smooth and level spot for the first 100 yards in front of the point from which it is intended to discharge these Rockets, as they generally travel in contact with the surface for this distance, not having acquired their full force, and are therefore more liable to deflection; but having at this point acquired a velocity not much less than the mean velocity of a cannon-ball, they are not to be more easily deflected: at this distance, also, they rise a few feet from the ground, so as to clear any ordinary obstacles that may occur; insomuch that, if it were desired to fire Rockets at low angles into a besieged town, from the third parallel, these Rockets, having a clear space to acquire their velocity, in front of the parallel, would run up the glacis, clear the ditch, and skim over the parapet into the town; and would no doubt be of great use in a variety of cases, particularly in discomfiting and rendering the enemy unsteady, by pouring in vollies of some hundreds or even thousands on this principle, previous to an assault or escalade: indeed, knowing the effect, I do not hesitate to affirm that this manœuvre, practised *on the great scale*, would infallibly dislodge any enemy posted for the protection of a breach.

Sufficient has, I conceive, now been stated, to give the officer such a general view of the power and spirit of the weapon, as may enable him to apply it in all possible cases to the best advantage; and if he will but constantly bear in view that maxim which I have laid down as the fundamental principle of this system, I will confidently pledge myself that it will never disappoint him, either as to the physical or *moral* effect which he may calculate on producing upon his enemy; since he must recollect, that for this latter effect, it adds all the terrors of *visibility* to every species of that destructive ammunition introduced by the use of gunpowder, but by every one admitted hitherto to have been qualified, as to moral effect, by its *invisibility*.

Note.—All the cases of service referred to in the above instructions will be found particularly detailed in the following plates.

STRENGTH OF A ROYAL ROCKET TROOP,

AS ORGANIZED IN THE BRITISH SERVICE IN 1822.

Captain.	Second Captain.	Lieutenants.	Staff Serjeants.	Serjeants.	Corporals.	Bombardiers.	Gunners.	Drivers.	Farrier.	Carriage Smith.	Shoeing Smiths.	Collar Makers.	Wheeler.	Trumpeter.	Total.	Horses.		
																Officers.	Troop.	Total.
1	1	2	2	3	3	7	97	36	1	1	2	2	1	1	160	12	233	245

Consisting of three Divisions, each Division comprising 10 Sections: *viz.*

- 10 Heavy, armed with 6-pounders, each Trooper carrying six in his holsters, and each section carrying one cavalry tube.
- 10 Medium, armed with 3-pounders, each Trooper carrying twelve, and each section one tube.
- 10 Light, armed with 1 and a half pounders, each Trooper carrying twenty-four, and each section two tubes.

18 Bat^{n}. Horses; *viz.:* six to each Division, carrying -

- 108 six-pounders.
- 216 three-pounders.
- 432 one-&-half-pounders.

COMPLEMENT OF CARRIAGES FOR THE TROOP.

One 18-pounder (4 Horses), with jointed Tube.
One 12-pounder (2 Horses), do.
Two 6-pounders (4 Horses), volley carriages.
Two 3-pounders (2 Horses), do.
Six Light Ammunition Waggons (4 Horses), for spare ammunition for the carriages.

The whole Troop, including the Mounted Sections, Battalion Horses, Carriages, and Light Ammunition Waggons, taking into the Field the following quantity of Ammunition, and number of Bouches à Feu :

	AMMUNITION.		AMMUNITION.
30 Mounted Sections carry	180 ... 6-pounders, 360 ... 3-pounders, 720 ... 1½-pounders,	Six Carriages carry	24 .. 18-pounders, 36 .. 12-pounders, 44 ... 6-pounders, 200 ... 3-pounders.
18 Bat^{n} Horses carry	100 .. 24-pounders, 108 ... 6-pounders, 216 ... 3-pounders, 432 ... 1½-pounders,	Six Light Ammunition Waggons carry	100 .. 18-pounders, 300 .. 12-pounders, 600 ... 6-pounders, 600 ... 3-pounders,

Bouches à Feu, 142.

Making a total of 4120 rounds of Rocket Ammunition of different calibres, and 142 Bouches à Feu, carried against the enemy; whereas a Troop of Horse Artillery, with the same establishment of Men, Horses, and Carriages, carries *only six* Bouches à Feu, and 1002 rounds of Ammunition into action.

(Signed) W. CONGREVE,
Major-General.

TABLE OF PLATES.

Plate 1

THE EQUIPMENT OF ROCKET CAVALRY.

PLATE 1. represents the mode of equipment for carrying Rockets on horseback, as it was arranged during the course of experiments, which were carried on under my direction, at Bagshot, in 1811; as it was subsequently carried into actual service, under Captain BOGUE, with the Allied armies in Germany, in the ever-memorable campaign of 1813; and as it is at present proposed to equip the new corps of Rocket Horse Artillery, established on the 1st of January 1814, by Earl MULGRAVE, Master General of the Ordnance, and composed of two troops, under the command of Lieutenant Colonel FISHER, of the Royal Artillery.

The right hand figure represents a trooper completely armed and equipped, in review order. The left hand figure is a delineation of the same, with the shabracque removed, to show the holsters in which the Rockets are conveyed. These holsters each contain two 6-pounder Rockets, each Rocket armed with a 6-pounder shell, or case shot; they are connected together at top, and are supported by the pummel of the saddle, which is made in the hussar fashion, though the saddle itself is, in fact, formed and stuffed the same as a common English saddle. This projection in front keeps the holsters clear of the horse's withers and shoulders, which, from their size, it might otherwise be difficult to do; for the latter of these purposes, also, the flap of the saddle comes further forward than usual. The holsters, thus connected, slip on and off from the pummel with great facility, which is an object of importance, as a part of the service of the Rocket trooper is, when from some impassable obstacle, he can no longer advance on horseback, to dismount and pass over such obstacle, with his ammunition holsters and chamber, on foot. The sticks, which are seven feet in length, and four in number, answering to the number of Rockets, are collected in a bundle by means of a strap with four loops, contrived on purpose, and are carried on the off-side, the thicker ends being supported in a bucket, suspended from the flap of the saddle, the strap above mentioned, as confining them together in the middle, leaning across the man's thigh to the peak of the saddle; by this means they fall naturally under his right arm, without at all incommoding him, either in mounting or dismounting, or even in going through the sword exercise. By this arrangement, also, they are easily drawn from the bundle downwards, for fixing to the Rocket, leaving any number that may remain as securely fixed as when the whole are in the quiver.

It has already been stated, that the men are told off in sections of threes. They are accordingly numbered 1, 2, and 3. Now Nos. 1. and 3. have nothing to carry but their proportion of the ammunition, viz. four Rockets and four sticks each, while No. 2., the left hand figure in the plate, has to carry [the tube or cylinder, from which the Rockets of his section are discharged. This tube, about five feet six inches in length, capable of being fixed steadily in the ground by four iron legs at the bottom of it, so that the Rockets may be discharged at low angles near the ground. The weight of this tube or bouche à feu, is about twelve lbs., and it is carried in a bucket similarly to the cavalry carbine.]

The men are armed with a sabre, which is in action suspended to the saddle, that they may not be incumbered in mounting and dismounting. Each man has besides a pistol in his cross belt, and a spear head in his holster, which may be occasionally fixed at the end of one of the Rocket sticks, so as to give the further aid of a very formidable lance. Instead of carrying slow match, which would be dangerous as well as inconvenient, the portfire is lighted in action by a flash of powder obtained from a pistol lock and pan, mounted on a small stock; and a light portfire stick for discharging the Rocket, about three feet in length, is constructed of a thin iron tube, which shuts up, and is carried in the holster. Every part of this equipment, except the sticks, is so completely concealed by the shabracque, that the Rocket trooper has the appearance merely of a lancer.

The weight of ammunition carried by the troop horse, with the full complement going into action, is less than two stone; to which the horse is fully equal for any ordinary operation. But in long marches, it would be not only useless but improvident to burthen him to this extent; small tumbrils, therefore, are provided to convey three rounds of each man's Rockets, he still carrying one round on the near side, and the four sticks on the off side to balance. The Rocket trooper has no heavy weights to lift — no guns to spunge, or to limber up and unlimber. He is required merely to be light and active for mounting and dismounting, and for moving nimbly on foot with a single Rocket, when in action: so that, whereas an artillery man cannot average less than 12 to more stone, the Rocket trooper need not exceed 10 stone, a difference amounting to the whole weight of ammunition carried by the men, even in action. It is needless to add, that this difference in the men must also give great facility in recruiting for a Rocket corps.

Plate 2

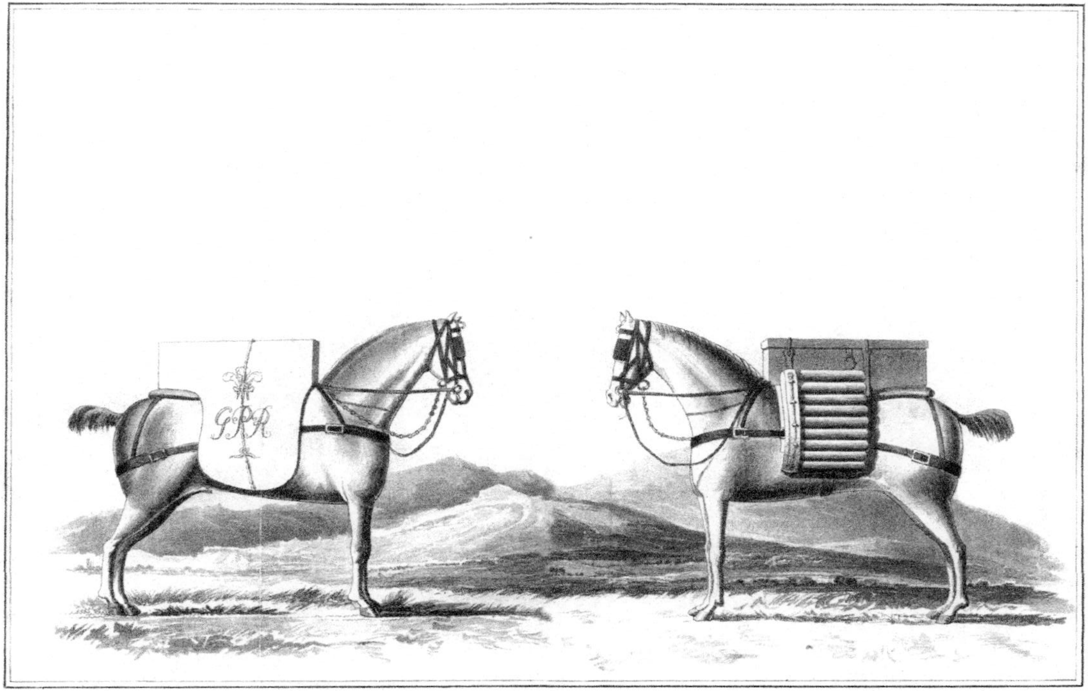

THE EQUIPMENT OF THE AMMUNITION HORSE.

PLATE 2. represents the mode of equipping the Ammunition Horses.

The left hand figure shows that the whole of the ammunition, &c. may be completely covered and protected from the weather by a painted canvass; and the other has this cover off, to show the particular distribution of the load, which consists of thirty-six 6-pounder Rockets and Rocket sticks, and a proportion of small stores, such as portfires, slow match, &c.

This load is carried on a bat saddle, made as small and as light as possible, with a pad at the back part of it, extending towards the crupper. The saddle is furnished on the top with two iron forks to receive a leather case, in which the sticks are carried.

The Rockets are carried in a sort of saddle bags, as they may be termed, stitched into separate compartments for each Rocket, covered by a flap at one end, and secured by a chain, staples, and padlocks, the Rocket lying horizontally. By this arrangement the load lies in the most compact form possible, and close to the horse's side, while the Rockets, being thus separated, cannot be injured by carriage.

The load is divided into three parts, the case or bundle of eighteen sticks, and a separate saddle bag on each side, contrived to hook on to the saddle, carrying eighteen Rockets in each bag. By this means there is no difficulty in loading and unloading the horse.

The whole weight thus carried by an Ammunition Horse is about 19 stone, consisting of about 6½ stone for the saddle, sticks, &c. and about six stone in each of the saddle bags. From which it is evident, that there is no fear of the load swagging the horse in travelling, because the centre of gravity is very considerably below his back bone. It is evident also, that as the weight of the Rockets diminishes by supplying the mounted men, the weight of the sticks also is diminished, and the centre of gravity may, if desired, be brought lower and lower, as the load diminishes, by taking the ammunition from the upper tiers gradually and equally on each side downwards. It is further evident, that although spaces are provided for eighteen Rockets in each bag, that number may be diminished, should the difficulty of the country, or the length of the march, or other circumstances, render it advisable to carry a less load.

The mode of leading these horses will be explained in the next Plate.

Fig. 1.

Fig. 2.

ROCKET CAVALRY IN LINE OF MARCH, AND IN ACTION.

Plate 3. Fig. 1. represents a sub-division of Rocket Cavalry, or Rocket Horse Artillery, marching in column of threes. It consists of six sections, of three men in each, or a less number of sections, according to the whole strength of the troop, followed by four ammunition horses, each pair led by a driver riding between them; on the full scale, therefore, a sub-division will consist of 24 horses and 20 men, and will carry into action 152 rounds of 6-pounder Shell or Case Shot Rockets, and six bouches à feu or tubes carried by the centre men of each section.

Fig. 2. represents this division in action, where the division may be supposed to have been halted in line, on the words — "*Prepare for action in front — dismount*" — Nos. 1. and 3. having dismounted, and given their leading reins to No. 2. who remains mounted, No. 1. runs forward about 15 or 20 paces with the tube, which he takes from the leather bucket in which it is slung from No. 2.'s saddle; and while Nos. 2. and 3. are preparing a Rocket, drawn from any one of the holsters most convenient, No. 1. fixes the tube by its legs firmly into the ground, pointing it to the desired object, and lights his portfire ready for the first round, which No. 3. by this time will have brought to him, and inserted into the rear end of the tube; there remains, then, only for No. 1. to touch the vent of the Rocket with his portfire, No. 3. having run back for another round, which No. 2. will have been able to prepare in the mean time. In this way the sub-division will, without hurry, come into action with six bouches a feu, in one minute's time, and may continue their fire, without any extraordinary exertion, at the rate of from two to three rounds from each tube in a minute, or even four with good exertion; so that the six bouches à feu would discharge 80 rounds of 6-pounder ammunition in three minutes.

The preparation of the Rocket for firing is merely by screwing the stick to it; great care, however, must be taken to fix the stick securely, as every thing depends on it; the vent also must be very carefully uncovered, as, if not perfectly so, the Rocket is liable to burst; and in firing, the portfire must not be thrust too far into the Rocket, for the same reason. In many cases sufficient accuracy may be given to the 6-pounder Rocket by sticks not exceeding 2 feet or 2 feet 6 inches in length; by which arrangement a certain number of rounds may be carried in the holsters ready fixed, the stick being carried downwards and passing through a hole at the bottom of the holster pipe.

On the words "*Cease firing*," No. 1. cuts his portfire, takes up his tube, runs back to his section, and replaces the tube immediately. No. 3. also immediately runs back; and having no other operation to perform, replaces the leading reins, and the whole are ready to mount again, for the performance of any further manœuvre that may be ordered, in less than a minute from the word "*Cease firing*" having been given.

It is obvious that the combined celerity and quantity of the discharge of ammunition of this description of artillery cannot be equalled or even approached, taking in view the means and nature of ammunition employed, by any other known system; the universality also of the operation, not being incumbered with wheel carriages, must be duly appreciated, as, in fact, it can proceed not only wherever cavalry can act, but even wherever infantry can get into action; it having been already mentioned that part of the exercise of these troops, supposing them to be stopped by walls, or ditches and morasses, impassable to horses, is to take the holsters and sticks from the horses, and advance on foot.

Another great advantage is the few men required to make a complete section, as by this means the number of points of fire is so greatly multiplied, compared to any other system of artillery. Thus it may be stated that the number of bouches à feu, which may comparatively be brought into action, by equal means, on the scale of a troop of horse artillery, would be at least six to one; and that they may either be spread over a great extent of line, or concentrated into a very small focus, according to the necessity of the service; indeed the skirmishing exercise of the Rocket Cavalry, divided and spread into separate sections, and returning by sound of bugle, forms a very interesting part of the system, and can be well imagined from the foregoing description and the annexed Plate.

Plate 1

Fig. 1.

Fig. 2.

ROCKET FIELD CARRIAGES IN LINE OF MARCH, AND IN ACTION.

PLATE 4. Fig. 1. represents a Rocket Carriage in line of march. There are various descriptions of these carriages. The general principle upon which they are constructed is that of multiplying the number of bouches à feu or cylinders for discharging the Rockets, in proportion to the nature of the Rocket to be used; so that each carriage, instead of being armed like a common piece of artillery with a single gun or cylinder, is provided with a number of these cylinders, which travel ready charged, so as to fire these Rockets in a volley on coming into action, or in rapid succession, as may be most expedient. Thus, a 24-pounder Rocket Carriage may be armed with four tubes, an 18-pounder with six, a 12-pounder with 8, a 9-pounder with ten, a 6-pounder with twelve or fourteen, and a 3-pounder with twenty; in addition to this, the carriage carries a large proportion of spare-ammunition in the limbers and store-boxes.

Fig. 2. represents these Rocket Carriages unlimbered for action: the one on the left hand gives a front view of a 12-pounder Volley Carriage, that on the right hand a side view of the same. The limbers are always supposed to be in the rear. The Rockets are fired with a portfire and long stick; and two men will fight the 3-pounder, 6-pounder, and 9-pounder Carriages, four men the 12-pounder, 18-pounder, and 24-pounder.

The exercise is very simple; on the word "*Prepare for action,*" the carriage being unlimbered, the front and rear flaps, which close the ends of the volley of tubes containing the Rockets, are opened at the same instant, by Nos. 1. and 2.; then taking hold of the end of the Rocket sticks contained in the tubes, draw back either the whole, or such number as it is intended to discharge in the volley, until the vent of the Rocket comes to the rear end of the tube, in which situation No. 3. discharges the Rockets either by drawing the portfire along the vents rapidly in volley, or in slower succession, No. 4. having pointed and elevated it.

Plate 5

Fig.1.

Fig.2

ROCKET INFANTRY IN LINE OF MARCH, AND IN ACTION.

PLATE 5. Fig. 1. represents a sub-division of Rocket infantry in line of march — Fig. 2. the same in action. The system here shown is the use of Rockets by infantry — one man in ten, or any greater proportion, carrying a Rocket tube complete, if it be not larger than the 3-pounder or 6-pounder, or two men carrying the same in half lengths, if it be the 9-pounder or 12-pounder. The construction of these tubes is similar to that of the cavalry tubes, standing upon four legs, but being longer and capable of greater elevation. The 9-pounder being extended in some cases to 15 feet, and the 12-pounder to 20 feet by which great accuracy is insured without any great incumbrance, as these tubes are divided into two or even three parts. The remainder of the party carry each from 3 to 6, according to the calibre of the Rocket; that is to say, two 12-pounders, three 9-pounders, four 6-pounders, eight 3-pounders. So that 100 men will bring into action, in any situation where musketry can be used, 200 rounds of 12-pounders, 300 rounds of 9-pounders, 400 of 6-pounders, 800 of 3-pounders.

The exercise and words of command are as follow:

No. 1. carries the light tube, or No. 1. and 2. the heavy, as above described; No. 1. also carries the pistol-portfire-lighter, and tube-box. No. 3. carries a small pouch, with the requisite small stores, such as spare tubes, portfires, &c., and a long portfire stick.

Nos. 4, 5, &c. to 10. carry each the Rockets as above described, each in a pouch on his back, sloping the sticks over his shoulder, secured together by straps and buckles.

With this distribution, they advance in double files. On the words, "*Halt,*" "*Prepare for action,*" being given, No. 1. places his tube, and with the assistance of No. 2. fixes it firmly into the ground in the required direction at the desired elevation. No. 2. then hands the portfire stick to No. 1. who prepares and lights it, while No. 2. steps back to receive the Rocket; which has been prepared by Nos. 3, 4, &c. who have fallen back on one side about fifteen paces, on the word being given to "*Prepare for action.*" These men can always supply the ammunition quicker than it can be fired, and one or other must therefore advance towards the frame to meet No. 2. with the round prepared. No. 2. having thus received the Rocket, inserts it into the rear end of the tube, then gives the word "*Ready,*" "*Fire,*" to No. 1. who discharges the Rocket. No. 2. then inserts, as before, another Rocket in the end of the tube, and stepping on one side, again gives the word "*Ready,*" "*Fire.*" By this process, from three to four Rockets a minute may, without difficulty, be fired from one frame, until the words "*Cease Firing,*" "*Prepare to advance,*" or "*Retreat,*" are given; when the frame is in a moment taken from the ground, and the whole party may either retire or advance immediately in press time, if required. To insure which, and at the same time to prevent any injury to the ammunition, Nos. 3, 4, &c. must not be allowed to take off their pouches, as they will be able to assist one another in preparing the ammunition, by only laying down their sticks; in taking up which again no time is lost.

If the tube is fired with a lock, the same process is used, except that No. 1. primes and cocks, and No. 2. fires on receiving the word from No. 1.

The Case-shot Rocket is an application of the arm peculiarly well adapted for the use of Cavalry or Infantry in the field. Every round of Case-shot Rockets contains a greater quantity of Carbine balls, than a round of Shrapnell's shells of corresponding calibre; the bursting powder of the Rocket Case-shot (the explosion of which is capable of the nicest regulation by fuse,) being contained in a chamber, in the rear of the charge of balls, as the Rocket passes through, the air gives an increased velocity to these balls on the bursting of the Case-shot, which is not the case with the Shrapnell Case-shot. In fact, the bursting powder of the Rocket Case-shot is so regulated, that the balls would penetrate an inch plank if the Rocket were at rest, and, consequently, however small the velocity of the Rocket may be at the time of the bursting of the Case-shot, the penetration of the balls will be always sufficient to produce the intended effect; whereas if the velocity of the Shrapnell shell is nearly expended, at the time that it bursts the balls fall harmless. The extraordinary powers of the Rocket in proportion to the means employed can in no way be better demonstrated, than by the effect produced by this description of Case-shot, against a screen, at 800 or a 1000 yards distance.

Fig. 1.

Fig. 2.

THE MODE OF USING ROCKETS IN BOMBARDMENT.

PLATE 6. Fig. 1. represents the mode of carrying the bombarding frame and ammunition by men. The apparatus required is merely a light pole about ten feet in length, having a 24-pounder or 32-pounder Rocket tube fixed by staples so as to take on and off at the upper end of it; the pole having two light legs attached to it, so as to support it like a triangle gin when fixed for bombarding. Every thing required for this service may be carried by men; or a Flanders-pattern ammunition waggon, with four horses, will convey sixty rounds of 32-pounder carcasses, in ten boxes, eight of the boxes lying crossways on the floor of the waggon, and two length-ways, at top. On these the bombarding frame is laid ready for use; and the sticks on each side, to complete the stowage of all that is necessary, the whole being covered by the tilt. Four men only are required to be attached to each waggon, who are numbered 1, 2, 3, and 4.

The frame and ammunition having been brought into the battery, or to any other place, concealed either by trees or houses (for from the facility of taking new ground, batteries are not so indispensable as with mortars), on the words "*Rear frame*" being given, Nos. 1. and 2. will, with ease, raise the frame, while 3. and 4. prepare the ammunition. Previous to charging the tube, the words "*Elevate to* 35°" or "45°," or whatever angle the officer may judge necessary, according to the required range, will be given, when the operation will be performed by Nos. 1. and 2., by spreading or closing the legs of the frame, agreeable to the distances marked in degrees on a small measuring tape, which the non-commissioned officer carries, and which is called — the Elevating Line. The word "*Point*" is then given; which is done by means of a plumb-line, hanging down from the vertex of the triangle, and which at the same time shows whether the frame is upright or not. Things being thus arranged, Nos. 1. and 2. adjust the front legs of the triangle until the tube is laid to the object. The word "*Load*" is then given, when No. 3. brings a Rocket and inserts it in the lower end of the tube, where it is retained in its position, the end of the stick resting on the ground, having before-hand *carefully* taken off the circle that covered the vent. If the Rocket is discharged by a lock,

No. 1. lays hold of the trigger-line, and retires ten or twelve paces obliquely, waiting for the word " *Fire.*"

On the Rocket leaving the frame, No. 1. receives another Rocket from No. 2., as before. In this routine any number of rounds is fired, until the words " *Cease firing*" are given; which, if followed by those, " *Prepare to retreat,*" Nos. 1. and 2. run forward to the ladder; and, on the words " *Lower frame,*" they ease it down in the same order in which it was raised, take it to pieces, and may thus retire in less than five minutes: or, if the object of ceasing to fire is merely a change of position to no great distance, the four men may with ease carry the frame, without taking it to pieces, the waggon following them with the ammunition, or the ammunition being borne by men, as circumstances may render expedient.

The ammunition projected from this frame consists of 32-pounder Rockets, armed with carcasses of the following sorts and ranges: —

1st. *The small carcass,* containing 8 lbs. of carcass composition, being 3 lbs. more than the present 10-inch spherical carcass. — Range 3,000 yards.

2d. *The medium carcass,* containing 12 lbs. of carcass composition, being equal to the present 13-inch. — Range 2,500 yards.

3d. *The large carcass,* containing 18 lbs. of carcass composition, being 6 lbs. more than the present 13-inch spherical carcass. — Range 2,000 yards.

Or, 32-pounder Rockets, armed with bursting cones, made of stout iron, filled with powder, to be exploded by fuzes, and to be used to produce the explosive effects of shells, where such effect is preferred to the conflagration of the carcass. These cones contain as follows: —

Small.—Five lbs. of powder, equal to the bursting powder of a 10-inch shell. — Range 3,000 yards.

Medium.—Eight lbs. of powder, equal to the bursting powder of a 13-inch shell. — Range 2,500 yards.

Large.—Twelve lbs. of powder. — Range 2,000 yards.

Plate 7

Fig. 1.

Fig. 2.

THE MODE OF USING ROCKETS IN BOMBARDMENT, FROM EARTH WORKS, WITHOUT APPARATUS.

PLATE 7. Fig. 1. is a perspective view of a Battery, erected expressly for throwing Rockets in bombardment, where the interior slope has the angle of projection required, and is equal to the length of the Rocket and stick.

The great advantage of this system is, that, as it dispenses with apparatus, where there is time for forming a work of this sort, of considerable length, the quantity of fire that may be thrown in a given time is limited only by the length of the work: thus, as the Rockets may be laid in embrasures cut in the bank, at every two feet, a battery of this description, 200 feet in length, will fire 100 Rockets in a volley, and so on: or an incessant and heavy fire may, by such a battery, be kept up from one flank to the other, and by replacing the Rockets as fast as they are fired in succession.

The rule for forming this battery is as follows:—

"The length of the interior slope of this work is half formed by the excavation, and half by the earth thrown out: for the base therefore of the interior slope of the part to be raised, at an angle of 55°, set off two thirds of the intended perpendicular height; cut down the slope to a perpendicular depth equal to the above mentioned height; then setting off, for the breadth of the interior excavation, one third more than the intended thickness of the work, carry down a regular ramp from the back part of this excavation to the foot of the slope, and the excavation will supply the quantity of earth necessary to give the exterior face a slope of 45°."

Fig. 2. is a perspective view of a common epaulement converted into a Rocket battery. In this case, as the epaulement is not of sufficient length to support the Rocket and stick, holes must be bored in the ground, with a miner's borer, of a sufficient depth to receive the sticks, and at such distances, and such an angle, as it is intended to place the Rockets for firing. The inside of the epaulement must be pared away to correspond with this angle, say 55°. The Rockets are then to be laid in embrasures, formed in the bank, as in the last case. Where the ground is such as to admit of using the borer; this latter system, of course, is the easiest operation; and for such ground as would be likely to crumble into the holes, slight tubes are provided, about two feet long, to preserve the opening; in fact, these tubes will be found advantageous in all ground.

Fig. 2. also shows a powerful mode of defending a field work by means of Rockets, in addition to the defences of the present system; merely by cutting embrasures in the glacis, for horizontal firing.

Fig. 1.

Fig. 2.

A ROCKET AMBUSCADE.

PLATE 8. Fig. 1. represents one of the most important uses that can be made of Rockets for field service; it is that of the Rocket Ambuscade for the defence of a pass, or for covering the retreat of an army, by placing any number, hundreds or thousands, of 32 or 24-pounder shell Rockets, or of 32-pounder Rockets, armed with 18-pounder shot, limited as to quantity only by the importance of the object which is to be obtained; as by this means the most extensive destruction, even amounting to annihilation, may be carried amongst the ranks of an advancing enemy, and that with the exposure of scarcely an individual.

The Rockets are laid in rows or batteries of 100 or 500 in a row, according to the extent of ground to be protected. They are to be concealed either in high grass, or masked in any other convenient way; and the ambuscade may be formed of any required number of these batteries, one behind the other, each battery being prepared to be discharged in a volley, by leaders of quick match: so that one man is, in fact, alone sufficient to fire the whole in succession, beginning with that nearest to the enemy, as soon as he shall have perceived them near enough to warrant his firing. Where the batteries are very extensive, each battery may be sub-divided into smaller parts, with separate trains to each, so that the whole, or any particular division of each battery, may be fired, according to the number and position of the enemy advancing. Trains, or leaders, are provided for this service, of a particular construction, being a sort of flannel saucissons, with two or three threads of slow match, which will strike laterally at all points, and are therefore very easy of application; requiring only to be passed from Rocket to Rocket, crossing the vents, by which arrangement the fire, running along from vent to vent, is sure to strike every Rocket in quick succession, without their disturbing each others' direction in going off, which they might otherwise do, being placed within 18 inches apart, if all were positively fired at the same instant.

Fig. 2. is a somewhat similar application, but not so much in the nature of an ambuscade as of an open defence. Here a very low work is thrown up, for the defence of a post, or of a chain of posts, consisting merely of as much earth and turf as is sufficient to form the sides of shallow embrasures for the large Rockets placed from two to three feet apart, or nearer; from which the Rockets are supposed to be discharged independently, by a certain number of artillery-men, employed to keep up the fire, according to the necessity of the case.

It is evident, that by this mode an incessant and tremendous fire may be maintained, which it would be next to impossible for an advancing enemy to pass through, not only from its quantity, and the weight and destructive nature of the ammunition, but from the closeness of its lines and its contiguity to the ground; leaving, in fact, no space in front which must not be passed over and ploughed up after very few rounds.

As both these operations are supposed to be employed in defensive warfare, and therefore in fixed stations, there is no difficulty involved in the establishment of a sufficient depôt of ammunition for carrying them on upon the most extensive scale; though it is obviously impossible to accomplish any thing approaching this system of defence by the ordinary means of artillery.

Fig. 1.

Fig. 2.

THE USE OF ROCKETS IN THE ATTACK AND DEFENCE OF FORTIFIED PLACES.

PLATE 9. Fig. 1. represents the advanced batteries and approaches in the attack of some fortress, where an imperfect breach being supposed to have been made in the salient angle of any bastion, large Rockets, weighing each from two to three hundred weight or more, and being each loaded with not less than a barrel of powder, are fired into the ruins after the revetment is broken, in order, by continual explosions, to render the breach practicable in the most expeditious way. To insure every Rocket that is fired having the desired effect, they are so heavily laden, as not to rise off the ground when fired along it; and under these circumstances are placed in a small shallow trench, run along to the foot of the glacis, from the nearest point of the third parallel, and in a direct line for the breach; by this means, the Rockets being laid in this trench will invariably pursue exactly the same course, and every one of them will be infallibly lodged in the breach. It is evident, that the whole of this is intended as a night operation, and a few hours would suffice, not only for running forward the trench, which need not be more than 18 inches deep, and about nine inches wide, undiscovered, but also for firing a sufficient number of Rockets to make a most complete breach before the enemy could take means to prevent the combinations of the operation.

From the experiments I have lately made, I have reason to believe, that Rockets much larger than those above mentioned may be formed for this description of service — Rockets from half a ton to a ton weight; which being driven in very strong and massive cast-iron cases, may possess such strength and force, that, being fired by a process similar to that above described, even against the revetment of any fortress, unimpaired by a cannonade, it shall, by its mass and form, pierce the same; and having pierced it, shall, with one explosion of several barrels of powder, blow such portion of the masonry into the ditch, as shall, with very few rounds, complete a practicable breach.

It is evident, from this view of the weapon, that the Rocket System is not only capable of a degree of portability, and facility for light movements, which no weapon possesses, but that its ponderous parts, or the individual masses of its ammunition, also greatly exceed those of ordinary artillery. And yet, although this last description of Rocket ammunition appears of an enormous mass as ammunition, still, if it be found capable of the powers here supposed, of which *I* have little doubt, the whole weight to be brought in this way against any town, for the accomplishment of a breach, will bear *no comparison* whatever to the weight of ammunition now required for the same service, independent of the saving of time and expence, and the great comparative simplicity of the approaches and works required for a siege carried on upon this system. This class of Rockets I propose to denominate the *Belier à feu.*

Fig. 2. represents the converse of this system, or the use of these larger Rockets for the defence of a fortress by the demolition of the batteries erected against it. In this case the Rockets are fired from embrasures, in the crest of the glacis, along trenches cut a part of the way in the direction of the works to be demolished.

Fig.1.

Fig.2.

OF THE USE OF ROCKETS BY INFANTRY AGAINST CAVALRY, AND IN COVERING THE STORMING OF A FORTRESS.

PLATE 10. Fig. 1. represents an attack of cavalry against infantry, repulsed by the use of Rockets. These Rockets are supposed to be of the lightest nature, 12 or 9-pounders, carried on bat horses or in small tumbrils, or with 6-pounder shell Rockets, of which one man is capable of carrying six in a bundle, for any peculiar service; or so arranged, that the flank companies of every regiment may be armed, each man, with such a Rocket, in addition to his carbine or rifle, the Rocket being contained in a small leather case, attached to his cartouch, slinging the carbine or rifle, and carrying the stick on his shoulder, serving him either as a spear, by being made to receive the bayonet, or as a rest for his piece.

By this means every battalion would possess a powerful battery of this ammunition, *in addition* to all its ordinary means of attack and defence, and with scarcely any additional burthen to the flank companies, the whole weight of the Rocket and stick not exceeding six pounds, and the difference between the weight of a rifle and that of a musket being about equivalent. As to the mode of using them in action, for firing at long ranges, as these Rockets are capable of a range of 2,000 yards, a few portable frames might be carried by each regiment, without any incumbrance, the frames for this description of Rocket not being heavier than a musket; but as the true intention of the arm, in this distribution of it, is principally for close quarters either in case of a charge of cavalry, or even of infantry, it is generally supposed to be fired in volleys, merely laid on the ground, as in the plate here described. And, as it is well known, how successfully charges of cavalry are frequently sustained by infantry, even by the fire of the musket alone, it is not presuming too much to infer, that the repulse of cavalry would be *absolutely certain*, by masses of infantry. possessing the additional aid of powerful volleys of these shell Rockets. So also in charges of infantry, whether the battalion so armed be about to charge, or to receive a charge, a well-timed volley of one or two hundred such

Rockets, judiciously thrown in by the flank companies, must produce the most decisive effects. Neither can it be doubted, that in advancing to an attack, the flank companies might make the most formidable use of this arm, mixed with the fire of their rifles or carbines, in all light infantry or tirailleur manœuvres. In like manner, in the passage of rivers, to protect the advanced party, or for the establishment of a *tête-du-pont*, and generally on all such occasions, Rockets will be found capable of the greatest service, as was shown, in the year 1813, in the passage of the Adour. In short, I must here remark that the use of the Rocket, in these branches of it, is no more limited than the use of gunpowder itself.

Fig. 2. represents the covering of the storm of a fortified place by means of Rockets. These are supposed to be of the heavy natures, both carcass and shell Rockets; the former fired in great quantities from the trenches at high angles; the latter in ground ranges in front of the third parallel. It cannot be doubted that the confusion created in any place, by a fire of some thousand Rockets, thus thrown in two or three volleys quickly repeated, must be most favourable, either to the storming of a particular breach, or to a general escalade.

I must here observe, that although, in all cases, I lay the greatest stress upon the use of this arm *in great quantities*, it is not therefore to be presumed, that the effect of an individual Rocket carcass, the smallest of which contains as much combustible matter as the 10-inch spherical carcass, is not at least equal to that of the 10-inch spherical carcass: or that the explosion of a shell thrown by a Rocket is not in its effects equal to the explosion of that same shell thrown by any other means: but that, as the power of *instantaneously* throwing the *most unlimited* quantities of carcasses or shells is the *exclusive property* of this weapon, and as there can be no question that an infinitely greater effect, both physical* as well as moral, is produced by the instantaneous application of any quantity of ammunition, with innumerable other advantages, than by a fire in slow succession of that same quantity: so it would be an absolute absurdity, and a downright waste of power, not to make this exclusive property the general basis of every application of the weapon, limited only by a due proportion between the expenditure and the value of the object to be attained—a limit which I should always conceive it more advisable to exceed than to fall short of.

There is another most important use in this weapon, in the storming of fortified places, which should here be mentioned, viz. that as it is the only

* For a hundred fires, breaking out at once, must necessarily produce more destruction than when they happen in succession, and may therefore be extinguished as fast as they occur.

description of artillery ammunition that can ever be carried into a place by a storming party, and as, in fact, the heaviest Rockets may accompany an escalade, so the value of it in these operations is infinite, and no escalade should ever be attempted without. It would enable the attackers, the moment they have got into the place, not only to scour the parapet most effectually, and to enfilade any street or passage where they may be opposed, and which they may wish to force; but even if thrown at random into the town, must distract the garrison, while it serves as a certain index to the different storming parties as to the situation and progress of each party.

THE USE OF ROCKETS FROM BOATS.

PLATE 11. represents two men of war's launches throwing Rockets. The frame is the same as that used for bombardment on shore, divested of the legs or prypoles, on which it is supported in land service; for which, afloat, the foremast of the boat is substituted. To render, therefore, the application of the common bombarding frame universal, each of them is constructed with a loop or traveller, to connect it with the mast, and guide it in lowering and raising, which is done by the haulyards.

The leading boat in the plate represents the act of firing; where the frame being elevated to any desired angle, the crew have retired into the stern-sheets, and a marine artillery-man is discharging a Rocket by a trigger-line leading aft. In the second boat, these artillery-men are in the act of loading; for which purpose, the frame is lowered to a convenient height; the mainmast is also standing, and the mainsail set, but partly brailed up. This sail being kept wet, most effectually prevents, without the least danger to the sail, any inconvenience to the men from the smoke or small sparks of the Rocket when going off; it should, therefore, be used where no objection exists on account of wind. It is not, however, by any means indispensable, as I myself discharged some hundred Rockets from these boats, nay, even from a six-oared cutter, without it. From this application of the sail, it is evident, that Rockets may be thrown from these boats under sail, as well as at anchor, or in rowing. In the launch, the ammunition may be very securely stowed in the stern-sheets, covered with tarpaulins, or tanned hides. In the six-oared cutter there is not room for this, and an attendant boat is therefore necessary: on which account, as well as from its greater steadiness, the launch is preferable, where there is no obstacle as to currents or shoal water.

Here it may be observed, with reference to its application in the marine, that as the power of discharging this ammunition without the burthen of ordnance, gives it *exclusive* facilities for land service, so also, its property of being projected without re-action upon the point of discharge, gives it *exclusive* facilities for sea service: insomuch, that Rockets conveying the same quantity of combustible matter, as by the ordinary system would be thrown from the largest mortars, and from ships of very heavy tonnage, may be used out of the smallest boats of the navy; and the 12-pounder and 18-pounder have been frequently fired even from four-oared gigs.

It should here also be remarked, that the 12 and 18-pounder shell Rockets recochet in the water remarkably well at low angles. There is another use for Rockets in boat service also, which ought not to be passed over; namely, their application in facilitating the capture of a ship by boarding.

In this service, 32-pounder shell Rockets are prepared with a short stick, having a leader and short fuze fixed to the stick for firing the Rocket. Thus prepared, every boat intended to board is provided with 10 or 12 of these Rockets; the moment of coming alongside, the fuzes are lighted, and the whole number of Rockets immediately launched by hand through the ports into the ship; where, being left to their own impulse, they will scour round and round the deck until they explode, so as very shortly to clear the way for the boarders, both by actual destruction, and by the equally powerful operation of terror amongst the crew; the boat lying quietly alongside for a few seconds, until, by the explosion of the Rockets, the boarders know that the desired effect has been produced, and that no mischief can happen to themselves when they enter the vessel.

Fig. 1.

Fig. 3.

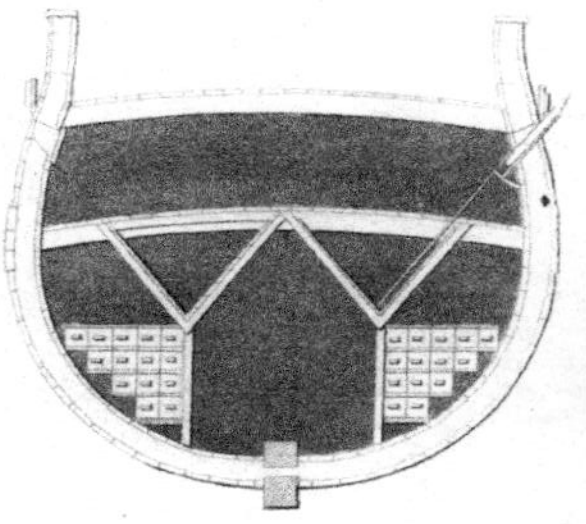

Fig. 2.

Fig. 4.

THE USE OF ROCKETS IN FIRE SHIPS, AND THE MODE OF FITTING ANY OTHER SHIP FOR THE DISCHARGE OF ROCKETS.

PLATE 12. Fig. 1. represents the application of Rockets in fire ships; by which, a great power of *distant* conflagration is given to these ships, in addition to the limited powers they now possess, as depending entirely on *contact* with the vessels they may be intended to destroy.

The application is made as follows: — Frames or racks are to be provided in the tops of all fire-ships, to contain as many hundred carcass and shell Rockets as can be stowed in them, tier above tier, and nearly close together. These racks may also be applied in the topmast and top-gallant shrouds, to increase the number: and when the time arrives for sending her against the enemy, the Rockets are placed in these racks, at different angles, and in all directions, having the vents uncovered, but requiring no leaders, or any nicety of operation, which can be frustrated either by wind or rain; as the Rockets are discharged merely by the progress of the flame ascending the rigging, at a considerable lapse of time after the ship is set on fire, and abandoned.

It is evident, therefore, in the first place, that no injury can happen to the persons charged with carrying in the vessel, as they will have returned into safety before any discharge takes place. It is evident, also, that the most extensive destruction to the enemy may be calculated on, as the discharge will commence about the time that the fire-ship has drifted in amongst the enemies' ships: when issuing in the most tremendous volleys, the smallest ship being supposed not to have less than 1000 Rockets, distributed in different directions, it is impossible but that every ship of the enemy must, with fire-ships enough, and no stint of Rockets, be covered sooner or later with clouds of this destructive fire; whereas, without this *distant power of destruction*, it is ten to one if every fire-ship does not pass harmlessly through the fleet, by the exertions of the enemies' boats in towing them clear—*exertions*, it must be remarked, *entirely precluded* in this system of fire-ships, as it is impossible that any boat could venture to approach a vessel so equipped, and pouring forth shell and carcass Rockets, in all directions, and at all angles. I had an opportunity of trying this experiment in the attack of the French fleet in Basque Roads, and though on a very small scale indeed, it was ascertained that the greatest confusion and terror was created by it in the enemy.

Figs. 2, 3, and 4. represent the mode of fitting any ship to fire Rockets, from scuttles in her broadside; giving thereby to every vessel having a between-deck a Rocket battery, in addition to the gun batteries on her spar deck, without the one interfering in the smallest degree with the other, or without the least risk to the ship; the sparks of the Rocket in going off being completely excluded, either by iron shutters closing the scuttle from within, as practised in the "Galgo" defence ship, fitted with 21 Rocket scuttles in her broadside, as shown in Fig. 3.; or by a particular construction of scuttle and frame which I have since devised, and applied to the Erebus sloop of war: so that the whole of the scuttle is completely filled, in all positions of traverse, and at all angles, by the frame; and thereby any possibility of the entrance of fire completely prevented. In both these ships, the Rockets may be either discharged at the highest angles, for bombardment, or used at low angles, as an additional means of offence or defence against other shipping in action; as the Rockets, thus used, are capable of projecting 18-pounder shot, or $4\frac{1}{2}$-inch shells, or even 24-pounder solid shot. This arrangement literally gives the description of small vessels here mentioned a second and most powerful deck for general service as well as for bombardment.

Smaller vessels, such as gun-brigs, schooners, and cutters, may be fitted to fire Rockets by frames, similar to the boat frames, described in Plate 11., from their spar deck, and either over the broadside or the stern; their frames being arranged to travel up and down, on a small upright spar or boat's mast, fixed perpendicularly to the outside of the bulwark of the vessel. As a temporary expedient, or in small vessels, this mode answers very well; but it has the objection of not carrying the sparks so far from the rigging, as when fired from below: it interferes also with the fighting the guns at the same time, and can therefore only be applied exclusively in the case of bombardment. All the gun-brigs, however, on the Boulogne station, during Commodore Owen's command there, were fitted in this manner, some with two and some with three frames on a broadside.

THE END.

London:
Printed by A. & R. Spottiswoode,
New-Street-Square.

Printed in Dunstable, United Kingdom

91218270R00060